T0331771

The Imperial College Lectures in

PETROLEUM ENGINEERING

Fluid Flow in Porous Media

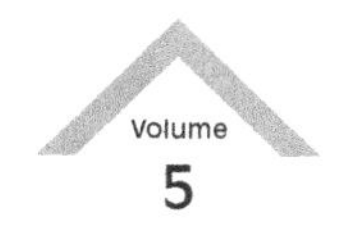

Other Related Titles from World Scientific

The Imperial College Lectures in Petroleum Engineering
Volume 1: An Introduction to Petroleum Geoscience
by Michael Ala
ISBN: 978-1-78634-206-5

The Imperial College Lectures in Petroleum Engineering
Volume 2: Reservoir Engineering
by Martin J Blunt
ISBN: 978-1-78634-209-6

The Imperial College Lectures in Petroleum Engineering
Volume 3: Topics in Reservoir Management
by Deryck Bond, Samuel Krevor, Ann Muggeridge, David Waldren and Robert Zimmerman
ISBN: 978-1-78634-284-3

The Imperial College Lectures in Petroleum Engineering
Volume 4: Drilling and Reservoir Appraisal
by Olivier Allain, Michael Dyson, Xudong Jing, Christopher Pentland, Marcel Polikar and Sander Suicmez
ISBN: 978-1-78634-395-6

The Imperial College Lectures in

PETROLEUM ENGINEERING

Fluid Flow in Porous Media

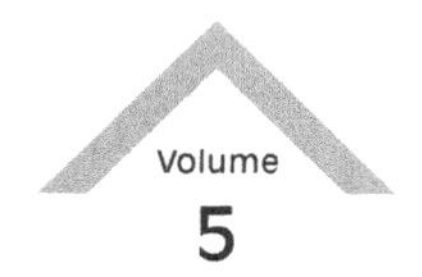

Robert W. Zimmerman
Imperial College London, UK

NEW JERSEY • LONDON • SINGAPORE • BEIJING • SHANGHAI • HONG KONG • TAIPEI • CHENNAI • TOKYO

Published by

World Scientific Publishing Europe Ltd.

57 Shelton Street, Covent Garden, London WC2H 9HE

Head office: 5 Toh Tuck Link, Singapore 596224

USA office: 27 Warren Street, Suite 401-402, Hackensack, NJ 07601

Library of Congress Cataloging-in-Publication Data
Names: Zimmerman, Robert Wayne, author.
Title: Fluid flow in porous media / by author: Robert Zimmerman
(Imperial College London, United Kingdom).
Description: [Hackensack] New Jersey : World Scientific, [2018] |
Series: The Imperial College lectures in petroleum engineering ; volume 5 |
Includes bibliographical references.
Identifiers: LCCN 2017048948 | ISBN 9781786344991 (hc : alk. paper)
Subjects: LCSH: Oil reservoir engineering. | Petroleum--Migration. | Fluids--Migration.
Classification: LCC TN870.57 .Z56 2018 | DDC 622/.3382--dc23
LC record available at https://lccn.loc.gov/2017048948

British Library Cataloguing-in-Publication Data
A catalogue record for this book is available from the British Library.

For any available supplementary material, please visit
http://www.worldscientific.com/worldscibooks/10.1142/Q0146#t=suppl

Desk Editors: Herbert Moses/Jennifer Brough/Shi Ying Koe

Typeset by Stallion Press
Email: enquiries@stallionpress.com

Printed in Singapore

Preface

This book is the fifth volume of a set of lecture notes based on the Master of Science course in Petroleum Engineering that is taught within the Department of Earth Science and Engineering at Imperial College London. The Petroleum Engineering MSc is a one-year course that comprises three components: (a) a set of lectures on the different topics that constitute the field of petroleum engineering, along with associated homework assignments and examinations; (b) a group field project in which the class is broken up into groups of about six students, who then use data from an actual reservoir to develop the field from the initial appraisal based on seismic and geological data, all the way through to eventual abandonment; and (c) a 14-week individual project, in which each student investigates a specific problem and writes a small "thesis" in the format of an SPE paper.

The Petroleum Engineering MSc course has been taught at Imperial College since 1976, and has trained over a thousand petroleum engineers. The course is essentially a "conversion course" that aims to take students who have an undergraduate degree in some area of engineering or physical science, but not necessarily any specific experience in petroleum engineering, and train them to the point at which they can enter the oil and gas industry as petroleum engineers. Although the incoming cohort has included students with undergraduate degrees in fields as varied as physics, mathematics, geology, and electrical engineering, the "typical" student on the course has an undergraduate degree in chemical or mechanical engineering, and little if any prior exposure to petroleum engineering. Although some students enter the course having had some experience in the oil industry, the course is intended to be self-contained, and prior knowledge of petroleum engineering or geology is not a prerequisite for any of the lecture modules.

The present volume presents the equations of fluid flow in porous media, with a focus on topics and issues that are relevant to petroleum reservoir engineering. No prior knowledge of this topic is assumed on the part of the reader, and particular attention is given to a careful mathematical and conceptual development of the governing equations, and to presenting solutions to important reservoir flow problems. The mathematical level is intended to be accessible to third- or fourth-year undergraduate students in engineering. Advanced topics such as Laplace transforms and Bessel functions, which play a key role in solving reservoir engineering problems, are developed in a self-contained manner. Although these notes have been written with a focus on petroleum engineering, it is expected that they will also be useful to hydrologists and civil engineers.

The author thanks the editorial and production staff at World Scientific for bringing this book to completion so rapidly and professionally. Special thanks go to Dr. Hanli de la Porte of PetroSim Consultants, who read the first draft of this book, and whose comments, criticisms, and suggestions helped to greatly improve its readability and practical relevance.

Robert W. Zimmerman

Imperial College London
January 2018

About the Author

Robert W. Zimmerman obtained a BS and MS in mechanical engineering from Columbia University, and a PhD in rock mechanics from the University of California at Berkeley. He has been a lecturer at UC Berkeley, a staff scientist at the Lawrence Berkeley National Laboratory, and Head of the Division of Engineering Geology and Geophysics at the Royal Institute of Technology (KTH) in Stockholm. He is the Editor-in-Chief of the *International Journal of Rock Mechanics and Mining Sciences*, and serves on the Editorial Boards of *Transport in Porous Media* and the *International Journal of Engineering Science*. He is the author of the monograph *Compressibility of Sandstones* (Elsevier, 1991), and co-author, with JC Jaeger and NGW Cook, of *Fundamentals of Rock Mechanics* (4th ed., Wiley-Blackwell, 2007). He is currently Professor of Rock Mechanics at Imperial College, where he conducts research on rock mechanics and fractured rock hydrology, with applications to petroleum engineering, underground mining, carbon sequestration, and radioactive waste disposal.

Contents

Chapter 1

Pressure Diffusion Equation for Fluid Flow in Porous Rocks

In this chapter, we will derive the basic differential equations that govern the time-dependent flow of fluids through porous media such as rocks or soils. This will be done by combining the principle of conservation of mass with Darcy's law, which relates the flow rate to the pressure gradient. The resulting differential equation will be a diffusion-type equation that governs the way that the fluid pressure changes as a function of time and varies spatially throughout the reservoir. The governing equations derived in this chapter form the basis of analytical models that are used in well test analysis, and, in discretised form, form the basis of numerical simulation codes that are used in petroleum reservoir engineering to predict oil and gas recovery.

1.1. Darcy's Law and the Definition of Permeability

The basic law governing the flow of fluids through porous media is Darcy's law, which was formulated by the French civil engineer Henry Darcy in 1856 on the basis of his experiments on vertical water filtration through sand beds. Darcy (1856) found that his data could be described by

$$Q = \frac{CA\Delta(P - \rho g z)}{L}, \tag{1.1.1}$$

where P is the pressure (Pa), ρ is the density (kg/m^3), g is the gravitational acceleration (m/s^2), z is the vertical coordinate (measured downwards) (m), L is the length of sample (m), Q is the volumetric flow rate (m^3/s), C is the constant of proportionality (m^2/Pa s), and A is the cross-sectional area of the sample (m^2).

Any consistent set of units can be used in Darcy's law, such as SI units, cgs units, British engineering units, etc. Unfortunately, in the oil and gas industry it is common to use so-called "oilfield units", which are not self-consistent. Oilfield units (i.e. barrels, feet, pounds per square inch, etc.) will not be used in these notes, except occasionally in some of the problems.

Darcy's law is mathematically analogous to other linear phenomenological transport laws, such as Ohm's law for electrical conduction, Fick's law for solute diffusion, and Fourier's law for heat conduction.

Why does the term "$P - \rho g z$" govern the flow rate? Recall from elementary fluid mechanics that Bernoulli's equation, which essentially embodies the principle of "conservation of energy", contains the terms

$$\frac{P}{\rho} - gz + \frac{v^2}{2} = \frac{1}{\rho}\left(P - \rho g z + \frac{\rho v^2}{2}\right), \qquad (1.1.2)$$

where P/ρ is related to the enthalpy per unit mass, $-gz$ is the gravitational energy per unit mass, and v^2/g is the kinetic energy per unit mass. Fluid velocities in a reservoir are usually very small, and so the third term is usually negligible, in which case we see that the combination $P - \rho g z$ represents an energy-type term. It seems reasonable that fluid would flow from regions of higher energy to lower energy, and, therefore, the driving force for flow should be the gradient (i.e. the rate of spatial change) of $P - \rho g z$.

Subsequent to Darcy's initial discovery, it has been found that, all other factors being equal, Q is inversely proportional to the fluid viscosity, μ (Pa s). It is therefore convenient to factor out μ, and put $C = k/\mu$, where k is known as the *permeability*, with dimensions (m^2).

It is usually more convenient to work with the volumetric flow per unit area, $q = Q/A$, rather than the total flow rate, Q. In terms of q, Darcy's law can be written as

$$q = \frac{Q}{A} = \frac{k}{\mu}\frac{\Delta(P - \rho g z)}{L}, \tag{1.1.3}$$

where the flux q has dimensions of (m/s). Since q is not quite the same as the velocity of the individual fluid particles (see Eq. (9.4.5)), it is perhaps better to think of these units as ($\mathrm{m}^3/\mathrm{m}^2\mathrm{s}$).

For transient processes in which the flux varies from point-to-point in a reservoir, we need a differential form of Darcy's law. In the vertical direction, this equation takes the form

$$q_v = \frac{Q}{A} = \frac{-k}{\mu}\frac{d(P - \rho g z)}{dz}, \tag{1.1.4}$$

where z is the downward-pointing vertical coordinate. The minus sign is included because the fluid flows in the direction from *higher* to *lower* values of $P - \rho g z$.

Since z is constant in the horizontal direction, the differential form of Darcy's law for one-dimensional (1D) horizontal flow is

$$q_H = \frac{Q}{A} = \frac{-k}{\mu}\frac{d(P - \rho g z)}{dx} = \frac{-k}{\mu}\frac{dP}{dx}. \tag{1.1.5}$$

For most sedimentary rocks, the permeability in the horizontal plane, k_H, is different than permeability in the vertical direction, k_V. Typically, $k_H > k_V$. The permeabilities in any two orthogonal directions within the horizontal plane may also differ. However, in these notes we will usually assume that $k_H = k_V$, and denote the permeability by k.

The permeability is a function of rock type, and also varies with stress, temperature, etc., but does not depend on the fluid; the effect of the fluid on the flow rate is accounted for by the viscosity term in Eq. (1.1.4) or (1.1.5).

Permeability has units of m^2, but in the petroleum industry it is conventional to use "Darcy" units, defined by

$$1 \text{ Darcy} = 0.987 \times 10^{-12}\,\mathrm{m}^2 \approx 10^{-12}\,\mathrm{m}^2. \tag{1.1.6}$$

The Darcy unit is defined such that a rock having a permeability of 1 Darcy would transmit 1 cm^3 of water (which has a viscosity of 1 cP) per sec, through a region of 1 cm^2 cross-sectional area, if the pressure drop along the direction of flow were equal to 1 atm per cm.

Different soils and sands that civil engineers deal with have permeabilities of a few Darcies. Hence, the original purpose of the "Darcy" unit was to avoid the need for using small prefixes such as 10^{-12}, as would be needed if k were measured in units of m^2. Fortunately, a Darcy is nearly a round number in SI units, so conversion between the two units is easy.

The numerical value of k for a given rock depends on the diameter of the pores in the rock, d, as well as on the degree of interconnectivity of the void space. Very roughly speaking, $k \approx d^2/1{,}000$; see Chapter 1 of Volume 3 of this series (Zimmerman, 2017a) for a derivation. For example, a rock with a typical pore size of 10×10^{-6} m, i.e. 10 microns, would be expected to have a permeability of about $10^{-13}\,\text{m}^2$, or 0.1 D. Typical values for intact (i.e. unfractured) rock are given in Table 1.1.

The permeabilities of different rocks and soils vary over many orders of magnitude. However, the permeabilities of petroleum reservoir rocks tend to be in the range of 0.001–1.0 Darcies. It is therefore convenient to quantify the permeability of reservoir rocks in units of a "milliDarcy" (mD), i.e. 0.001 D.

Table 1.1. Range of values of permeabilities of various rock types.

Rock type	k (Darcies)	k (m^2)
Coarse gravel	10^{3}–10^{4}	10^{-9}–10^{-8}
Sands, gravels	10^{0}–10^{3}	10^{-12}–10^{-9}
Fine sand, silt	10^{-4}–10^{0}	10^{-16}–10^{-12}
Clay, shales	10^{-9}–10^{-6}	10^{-21}–10^{-18}
Limestones	10^{-4}–10^{0}	10^{-16}–10^{-12}
Sandstones	10^{-5}–10^{1}	10^{-17}–10^{-11}
Weathered chalk	10^{0}–10^{2}	10^{-12}–10^{-10}
Unweathered chalk	10^{-9}–10^{-1}	10^{-21}–10^{-13}
Granite, gneiss	10^{-8}–10^{-4}	10^{-20}–10^{-16}

The values in Table 1.1 are for intact rock. In some reservoirs, the permeability is mainly due to an interconnected network of fractures. The permeabilities of fractured rock masses tend to be in the range of 1 mD–10 Darcies. In a fractured reservoir, the reservoir-scale permeability is not directly related to the core-scale permeability that one would measure in the laboratory on an unfractured core.

1.2. Datum Levels and Corrected Pressure

If the fluid is in static equilibrium, then $q = 0$, and Eq. (1.1.4) yields

$$\frac{d(P - \rho g z)}{dz} = 0 \rightarrow P - \rho g z = \text{constant}. \tag{1.2.1}$$

If we take $z = 0$ to be at the surface, where the fluid pressure is atmospheric, then the static reservoir fluid pressure at a depth z is

$$P_{\text{static}}(z) = P_{\text{atm}} + \rho g z. \tag{1.2.2}$$

As we always measure the reservoir pressure as "gauge pressure" (i.e. the pressure above atmospheric, as measured by a downhole pressure gauge), we can essentially neglect P_{atm} in Eq. (1.2.2). We then see by comparing Eq. (1.2.2) with Eq. (1.1.4) that only the pressure above and beyond the static pressure given in Eq. (1.2.2) plays a role in "driving" the flow. In a sense, then, the term $\rho g z$ is superfluous because it only contributes to the static pressure, but does not contribute to the driving force for flow.

In order to remove this extraneous term, it is common to define a corrected pressure, P_c, as

$$P_c = P - \rho g z. \tag{1.2.3}$$

(Note that this "corrected pressure" is unrelated to the capillary pressure, defined in Section 1.8, which is also usually denoted by P_c.) In terms of the corrected pressure, Darcy's law (for, say, horizontal flow) can be written as

$$q = \frac{Q}{A} = \frac{-k}{\mu}\frac{dP_c}{dx}. \tag{1.2.4}$$

Instead of using the surface level ($z = 0$) as the "datum", we often use some depth z_o such that equal amounts of initial oil-in-place lie above and below z_o. In this case,

$$P_c = P - \rho g(z - z_o). \tag{1.2.5}$$

The choice of the datum level is immaterial, in the sense that it only contributes a constant term to the corrected pressure, and so does not contribute to the pressure gradient. The corrected pressure defined in Eq. (1.2.5) can be interpreted as the pressure of a hypothetical fluid at depth z_o that would be in hydrostatic equilibrium with the fluid that exists at the actual pressure at depth z.

1.3. Concept of Representative Elementary Volume

Darcy's law is a macroscopic law that is intended to be meaningful over regions that are much larger than the size of a single pore. In other words, when we talk about the permeability at a point "(x, y, z)" in the reservoir, we cannot be referring to the permeability at a mathematically infinitesimal "point" because a given point may, for example, lie in a sand grain, not in the pore space!

The property of permeability is in fact only defined for a porous medium, not for an individual pore. Hence, the permeability is a property that is in some sense "averaged out" over a certain region of space surrounded by the mathematical point (x, y, z). This region must be large enough to encompass a statistically significant number of pores. Likewise, the "pressure" that we use in Darcy's law is actually an average pressure taken over a small region of space.

For example, consider Figure 1.1, which shows a few pores in a sandstone. Two position vectors, R_1 and R_2, are indicated in the figure. However, when we refer to the "pressure" at a certain location in the reservoir, we do not distinguish between two nearby points such as these. Instead, the entire region shown in the figure would be represented by an average pressure that is taken over the indicated circular region, which is known as a "representative elementary volume" (REV). Similarly, the permeability of a rock is only defined over the REV length scale (Bear, 1972).

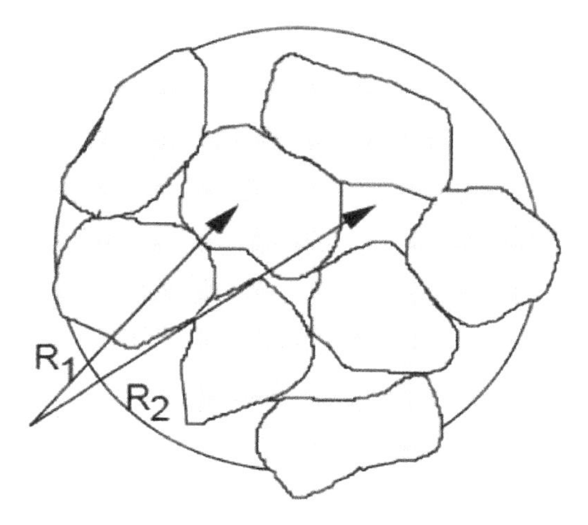

Figure 1.1. Representative elementary volume.

Roughly, in sandstones, the REV scale must be at least one order of magnitude larger than the mean pore size. However, in heterogeneous rocks such as carbonates, the pore size may vary spatially in an irregular manner, and an REV might not exist. Although it is important to be aware of this concept, for most reservoir engineering purposes, no explicit consideration of this issue is required.

1.4. Radial, Steady-state Flow to a Well

Before we derive the general transient equation that governs fluid flow through porous media, we will examine a simple, but illustrative, problem that can be solved using only Darcy's law: a circular reservoir that has a constant pressure at its outer boundary, and a constant flow rate into the wellbore.

Consider (see Figure 1.2) a reservoir of thickness H and horizontal permeability k, fully penetrated by a vertical well of radius R_w. Assume that at some radius R_o, the pressure remains at its undisturbed value, P_o. If we pump oil from this well at a rate Q, what will be the steady-state pressure distribution in the reservoir?

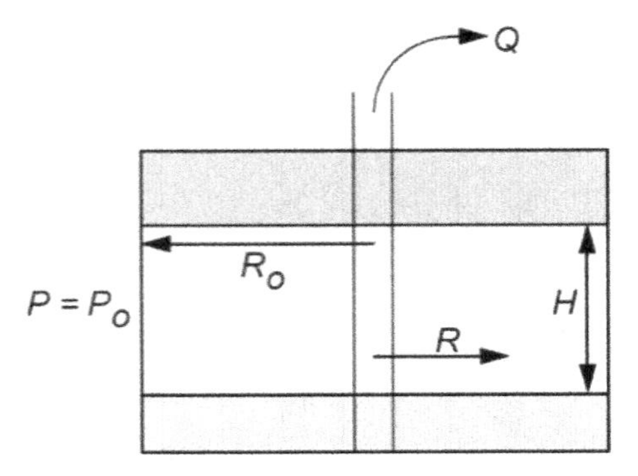

Figure 1.2. Well in a bounded circular reservoir (side view).

Note that, initially, the pressure at any location in the reservoir will vary with time; this transient problem is discussed and solved in Section 6.3. But eventually, the pressure distribution in this reservoir will reach a steady state, in which $dP/dt = 0$. This is the situation that we will now consider. In the steady state, the flow rate into the reservoir at $R = R_o$ will exactly equal the flow rate into the wellbore at $R = R_w$.

The analogue of Eq. (1.1.5) for flow in the R direction is

$$Q = \frac{-kA}{\mu}\frac{dP}{dR}. \tag{1.4.1}$$

The cross-sectional area normal to the flow, at a radial distance R from the centre of the well, is $2\pi RH$ (i.e. a cylindrical surface of height H and perimeter $2\pi R$), so

$$Q = \frac{-2\pi kH}{\mu}R\frac{dP}{dR}. \tag{1.4.2}$$

Separate the variables, and integrate from the outer boundary, $R = R_o$, to some generic location R:

$$\frac{dR}{R} = \frac{-2\pi kH}{\mu Q}dP,$$

$$\Rightarrow \int_{R_o}^{R}\frac{dR}{R} = -\int_{P_o}^{P}\frac{2\pi kH}{\mu Q}dP,$$

$$\Rightarrow \ln\left(\frac{R}{R_o}\right) = \frac{-2\pi kH}{\mu Q}(P - P_o),$$

$$\Rightarrow P(R) = P_o - \frac{\mu Q}{2\pi kH}\ln\left(\frac{R}{R_o}\right). \tag{1.4.3}$$

Equation (1.4.3) is the famous Dupuit–Thiem equation, first derived in 1857 by the French hydrologist Jules Dupuit (Dupuit, 1857). The German hydrologist Adolf Thiem (Thiem, 1887) seems to have been the first to popularise the use of this equation to estimate the permeability of groundwater aquifers.

Since the pressure varies logarithmically with distance from the wellbore (Figure 1.3), most of the drawdown occurs near the well, whereas far from the well, the pressure varies slowly.

We can make the following comments about Eq. (1.4.3):

- If fluid is pumped from the well, then (mathematically) Q is negative because the fluid is flowing in the direction opposite to the direction of the radial coordinate, R. Hence, $P(R)$ will be less than P_o for any $R < R_o$.
- The amount by which $P(R)$ is less than P_o is called the pressure drawdown.
- The only reservoir parameter that affects the pressure drawdown is the "permeability-thickness" product, kH.

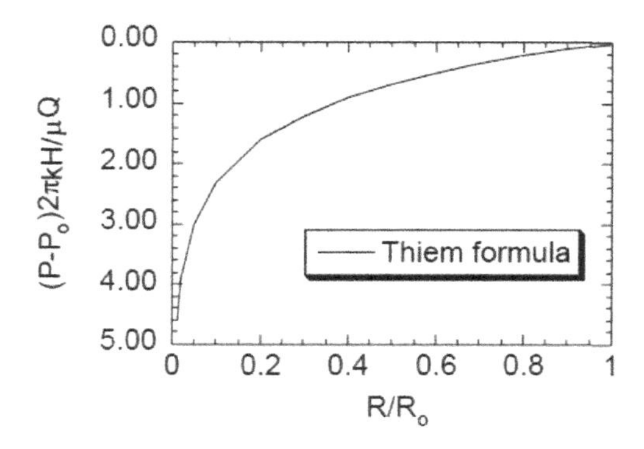

Figure 1.3. Steady-state flow to a well in a circular reservoir.

- The pressure varies logarithmically as a function of radial distance from the well. This same type of dependence also occurs in transient problems, as will be seen in Chapter 2.
- The pressure drawdown at the well is found by setting $R = R_w$ in Eq. (1.4.3):

$$P_w = P_o - \frac{\mu Q}{2\pi k H} \ln\left(\frac{R_w}{R_o}\right). \tag{1.4.4}$$

- Since we are usually interested in situations in which the fluid is flowing towards the well (i.e. "production"), it is common to redefine Q to be positive for production, in which case we write Eq. (1.4.4) as

$$P_w = P_o + \frac{\mu Q}{2\pi k H} \ln\left(\frac{R_w}{R_o}\right). \tag{1.4.5}$$

1.5. Conservation of Mass Equation

Darcy's law in itself does not contain sufficient information to allow us to solve transient (i.e. time-dependent) problems involving subsurface flow. In order to develop a complete governing equation that applies to transient problems, we must first derive a mathematical expression of the principle of conservation of mass.

Consider flow through a 1D tube of cross-sectional area A; in particular, let us focus on the region between two locations x and $x + \Delta x$, as in Figure 1.4. The main idea behind the application of the principle of conservation of mass is

$$\text{Flux in} - \text{Flux out} = \text{Increase in amount stored}. \tag{1.5.1}$$

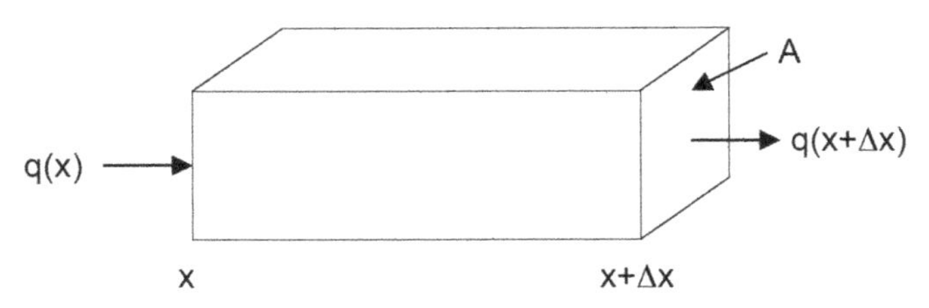

Figure 1.4. Prismatic region used to derive the equation for conservation of mass.

Note that the property that is conserved is the *mass* of the fluid, not the *volume* of the fluid.

Consider the period of time between time t and time $t + \Delta t$. To be concrete, assume that the fluid is flowing from left to right through the core. During this time increment, the mass flux into this region of rock between will be

$$\text{Mass flux in} = A(x)\rho(x)q(x)\Delta t. \tag{1.5.2}$$

The mass flux out of this region of rock will be

$$\text{Mass flux out} = A(x + \Delta x)\rho(x + \Delta x)q(x + \Delta x)\Delta t. \tag{1.5.3}$$

The amount of fluid mass stored in the region is denoted by m, so the conservation of mass equation takes the form

$$\begin{aligned}[A(x)\rho(x)q(x) - A(x + \Delta x)\rho(x + \Delta x)q(x + \Delta x)]\Delta t \\ = m(t + \Delta t) - m(t).\end{aligned} \tag{1.5.4}$$

For 1D flow, such as through a cylindrical core, $A(x) = A = \text{constant}$. In this case, we can factor out A, divide both sides by Δt, and let $\Delta t \to 0$

$$-A[\rho q(x + \Delta x) - \rho q(x)] = \lim_{\Delta t \to 0} \frac{m(t + \Delta t) - m(t)}{\Delta t} = \frac{\partial m}{\partial t}, \tag{1.5.5}$$

where we temporarily treat the product ρq as a single entity.

But $m = \rho V_p$, where V_p is the pore volume of the rock contained in the slab between x and $x + \Delta x$. The pore volume is, by definition, equal to the macroscopic volume multiplied by the porosity, ϕ. So

$$m = \rho V_p = \rho \phi V = \rho \phi A \Delta x, \tag{1.5.6}$$

$$\Rightarrow -A[\rho q(x + \Delta x) - \rho q(x)] = \frac{\partial(\rho\phi)}{\partial t} A \Delta x. \tag{1.5.7}$$

Now divide both sides by $A\Delta x$, and let $\Delta x \to 0$

$$-\frac{\partial(\rho q)}{\partial x} = \frac{\partial(\rho\phi)}{\partial t}. \tag{1.5.8}$$

Equation (1.5.8) is the basic equation of conservation of mass for 1D linear flow in a porous medium, which relates the spatial rate of change of stored mass to the temporal rate of change of stored mass. It is exact, and applies to gases, liquids, high or low flow rates, etc.

In its most general, 3D form, the equation of conservation of mass can be written as

$$\frac{\partial(\rho q)}{\partial x} + \frac{\partial(\rho q)}{\partial y} + \frac{\partial(\rho q)}{\partial z} = -\frac{\partial(\rho\phi)}{\partial t}. \tag{1.5.9}$$

The mathematical operation on the left-hand side of Eq. (1.5.9) is known as the *divergence* of ρq, which represents the rate at which fluid diverges from a given region, per unit volume.

1.6. Diffusion Equation in Cartesian Coordinates

True steady-state flow rarely occurs in an oil or gas reservoir. The typical flow scenario is a transient one, in which the pressure, density, flow rate, etc., all vary in space and time. Transient flow of a fluid through a porous medium is governed by a type of partial differential equation known as a diffusion equation. Although mathematically analogous to the diffusion-type equations that govern the flow of heat through a solid body, or the diffusion of solute particles through a liquid, for example, it may be worth noting that, unlike those processes, pressure diffusion in a reservoir is not driven by any underlying stochastic or random process at the molecular scale, but is simply driven by Darcy's law.

In order to derive the pressure diffusion equation, we combine Darcy's law, the conservation of mass equation, and an equation that relates the pore fluid pressure to the amount of fluid that is stored inside the porous rock. (Strangely enough, this last aspect of flow through porous media only came to be understood many decades after Darcy's law was discovered!)

Let us look more closely at the right-hand side of Eq. (1.5.8), and use the product rule, and chain rule, of differentiation

$$\begin{aligned}\frac{\partial(\rho\phi)}{\partial t} &= \rho\frac{\partial\phi}{\partial t} + \phi\frac{\partial\rho}{\partial t}\\ &= \rho\frac{d\phi}{dP}\frac{\partial P}{\partial t} + \phi\frac{d\rho}{dP}\frac{\partial P}{\partial t}\\ &= \rho\phi\left[\left(\frac{1}{\phi}\frac{d\phi}{dP}\right) + \left(\frac{1}{\rho}\frac{d\rho}{dP}\right)\right]\frac{\partial P}{\partial t}\\ &= \rho\phi(c_\phi + c_f)\frac{\partial P}{\partial t},\end{aligned} \tag{1.6.1}$$

where c_f is the compressibility of the fluid, and c_ϕ is the "pore compressibility" of the rock formation, also sometimes called the "formation compressibility" (Matthews and Russell, 1967).

Note that the above derivation, which is traditional in petroleum engineering, implicitly assumes that the porosity may change, but the reservoir itself is macroscopically rigid. This is of course physically inconsistent. However, a more rigorous derivation would require consideration of rock deformation, and lies outside the scope of these notes. Such a derivation can be found in de Marsily (1986). Furthermore, since reservoirs compact vertically when fluid is extracted from them, it must be noted that the pore compressibility term must be calculated under conditions of uniaxial (i.e. vertical) strain (Zimmerman, 2017b).

Now look at the left-hand side of Eq. (1.5.8). The volumetric flow per unit area, q, is given by Darcy's law, Eq. (1.1.5), and so the left-hand side of Eq. (1.5.8) becomes

$$\begin{aligned}-\frac{\partial(\rho q)}{\partial x} &= -\frac{\partial}{\partial x}\left[\frac{-\rho k}{\mu}\frac{\partial P}{\partial x}\right] = \frac{k}{\mu}\left[\rho\frac{\partial^2 P}{\partial x^2} + \frac{\partial\rho}{\partial x}\frac{\partial P}{\partial x}\right]\\ &= \frac{k}{\mu}\left[\rho\frac{\partial^2 P}{\partial x^2} + \frac{d\rho}{dP}\frac{\partial P}{\partial x}\frac{\partial P}{\partial x}\right]\end{aligned}$$

$$= \frac{\rho k}{\mu}\left[\frac{\partial^2 P}{\partial x^2} + \left(\frac{1}{\rho}\frac{d\rho}{dP}\right)\left(\frac{\partial P}{\partial x}\right)^2\right]$$

$$= \frac{\rho k}{\mu}\left[\frac{\partial^2 P}{\partial x^2} + c_f\left(\frac{\partial P}{\partial x}\right)^2\right]. \tag{1.6.2}$$

Now equate Eqs. (1.6.1) and (1.6.2) to arrive at

$$\frac{\partial^2 P}{\partial x^2} + c_f\left(\frac{\partial P}{\partial x}\right)^2 = \frac{\phi\mu(c_f + c_\phi)}{k}\frac{\partial P}{\partial t}. \tag{1.6.3}$$

For liquids, the second term on the left is negligible compared to the first. To justify this assertion, we can use Eq. (1.4.5), and ignore the difference between x and R, to find

$$c_f\left(\frac{\partial P}{\partial x}\right)^2 \approx c_f\left(\frac{\mu Q}{2\pi kHR}\right)^2, \tag{1.6.4}$$

$$\frac{\partial^2 P}{\partial x^2} \approx \frac{\mu Q}{2\pi kHR^2}, \tag{1.6.5}$$

$$\Rightarrow \text{Ratio} = \frac{c_f \mu Q}{2\pi kH} = \frac{c_f(P_o - P_w)}{\ln(R_o/R_w)}. \tag{1.6.6}$$

Typical values of these parameters, for liquids, are

$$c_f \approx 10^{-10}/\text{Pa},$$

$$P_o - P_w \approx 10 \text{ MPa}{=}10^7 \text{ Pa},$$

$$\ln(R_o/R_w) \approx \ln(1000\,\text{m}/0.1\,\text{m}) = \ln(10^4) \approx 10,$$

$$\Rightarrow \text{Ratio} = \frac{10^{-10} \times 10^7}{10} = 10^{-4} \ll 1. \tag{1.6.7}$$

This example shows that, for liquids, the nonlinear term in Eq. (1.6.3) is small, and so is always neglected, in practice. For gases, however, this term cannot be neglected (see Chapter 9).

Table 1.2. Typical values of the compressibility of various rock types and reservoir fluids.

Rock (or fluid) type	c (1/Pa)	c (1/psi)
Sand	10^{-6}–10^{-8}	10^{-2}–10^{-4}
Sandstones	10^{-7}–10^{-9}	10^{-3}–10^{-5}
Carbonates	10^{-9}–10^{-11}	10^{-9}–10^{-11}
Shales	10^{-10}–10^{-12}	10^{-10}–10^{-12}
Water	5×10^{-10}	3.5×10^{-6}
Oil	1×10^{-9}	7.0×10^{-6}

The 1D, linearised form of the pressure diffusion equation is therefore

$$\frac{\partial P}{\partial t} = \frac{k}{\phi \mu c_t} \frac{\partial^2 P}{\partial x^2}, \tag{1.6.8}$$

in which the total compressibility is given by

$$c_t = c_{\text{formation}} + c_{\text{fluid}} = c_\phi + c_f. \tag{1.6.9}$$

The product of the compressibility and the porosity, ϕc_t, is called the storativity. Typical ranges of the values of these compressibilities are shown in Table 1.2.

For some rocks, the pore compressibility is negligible compared to the fluid compressibility, and the storativity is mainly due to the fluid compressibility. For soils and unconsolidated sands, the opposite is often the case. In general, both contributions to the total compressibility must be taken into account.

Much of the remainder of these notes will be devoted to solving the diffusion equation in various situations. For now, we make the following general remarks about it:

- The parameter that governs the rate at which fluid pressure diffuses through a porous rock mass is the hydraulic diffusivity (m^2/s), which is defined by

$$D_H = \frac{k}{\phi \mu c_t}. \tag{1.6.10}$$

- Roughly speaking (this will be proven in Section 2.5), the distance R over which a pressure disturbance travels during an elapsed time t is

$$R = \sqrt{4D_H t} = \sqrt{\frac{4kt}{\phi\mu c_t}}. \tag{1.6.11}$$

As an example, in a reservoir having $\phi = 0.2$, $k = 100$ mD, $c_t = 10^{-9}$/Pa, and $\mu = 1$ cP, the radius of penetration will be about 85 m after fluid has been flowing to or from a well for one hour.
- Conversely, the time required for a pressure disturbance to travel a distance R from the well into the reservoir is found by inverting Eq. (1.6.11)

$$t = \frac{\phi\mu c_t R^2}{4k}. \tag{1.6.12}$$

- Pressure pulses obey a diffusion equation, not a wave equation, such as governs the propagation of seismic waves, for example. Rather than travelling at a constant speed, the pressure pulse propagates at a speed that continually decreases with time. To prove this, differentiate Eq. (1.6.11) with respect to time, and observe that the "velocity" of the pulse, dR/dt, decays like $1/\sqrt{t}$.

1.7. Diffusion Equation in Cylindrical Coordinates

In petroleum engineering, we are usually interested in fluid flowing towards a well, in which case it is more convenient to use cylindrical (radial) coordinates, rather than Cartesian coordinates.

To derive the proper form of the diffusion equation in radial coordinates, consider fluid flowing towards, or away from, a vertical well in a homogeneous reservoir of uniform thickness H, in a radially symmetric manner. We now perform a mass balance on a thin annular region between R and $R + \Delta R$ (Figure 1.5). Returning to Eq. (1.5.4), we replace x with R, and note that for the annular

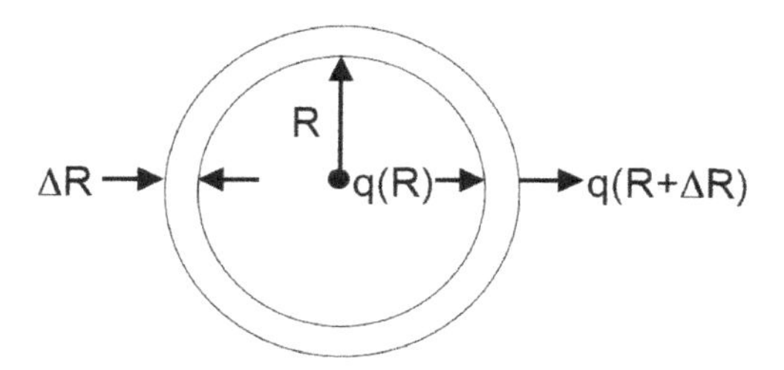

Figure 1.5. Annular region used to derive the pressure diffusion equation in radial coordinates.

geometry, $A(R) = 2\pi RH$, thus obtaining

$$\begin{aligned}[2\pi RH\rho(R)q(R) - 2\pi(R+\Delta R)H\rho(R+\Delta R)q(R+\Delta R)]\Delta t \\ = m(t+\Delta t) - m(t). \qquad (1.7.1)\end{aligned}$$

As before, divide by Δt, and let $\Delta t \to 0$:

$$2\pi H[R\rho(R)q(R) - (R+\Delta R)\rho(R+\Delta R)q(R+\Delta R)] = \frac{\partial m}{\partial t}. \qquad (1.7.2)$$

On the right-hand side,

$$m = \rho\phi V = \rho\phi 2\pi HR\Delta R, \qquad (1.7.3)$$

$$\Rightarrow \frac{\partial m}{\partial t} = \frac{\partial(\rho\phi 2\pi HR\Delta R)}{\partial t} = 2\pi HR\frac{\partial(\rho\phi)}{\partial t}\Delta R. \qquad (1.7.4)$$

Equate Eqs. (1.7.2) and (1.7.4), divide by ΔR, and let $\Delta R \to 0$:

$$-\frac{\partial(\rho qR)}{\partial R} = R\frac{\partial(\rho\phi)}{\partial t}. \qquad (1.7.5)$$

Equation (1.7.5) is the radial-flow version of the conservation of mass equation, which in the linear case was given by Eq. (1.5.8).

Now use Darcy's law in the form of Eq. (1.4.1) for q on the left-hand side, and insert Eq. (1.6.1) on the right-hand side:

$$\frac{k}{\mu}\frac{\partial}{\partial R}\left(\rho R\frac{\partial P}{\partial R}\right) = \rho\phi(c_f + c_\phi)R\frac{\partial P}{\partial t}. \qquad (1.7.6)$$

Following the same procedure as that which led to Eq. (1.6.3), we find

$$\frac{1}{R}\frac{\partial}{\partial R}\left(R\frac{\partial P}{\partial R}\right) + c_f\left(\frac{\partial P}{\partial R}\right)^2 = \frac{\phi\mu(c_f + c_\phi)}{k}\frac{\partial P}{\partial t}. \tag{1.7.7}$$

For liquids, we again neglect the term $c_f(\partial P/\partial R)^2$, to arrive at

$$\frac{\partial P}{\partial t} = \frac{k}{\phi\mu c_t}\frac{1}{R}\frac{\partial}{\partial R}\left(R\frac{\partial P}{\partial R}\right). \tag{1.7.8}$$

Equation (1.7.8) is the governing equation for transient, radial flow of a liquid through porous rock. It is the governing equation for flow during primary production, and is the starting point for well test analysis methods. We will develop and analyse solutions to this equation in later parts of these notes.

1.8. Governing Equations for Multi-phase Flow

In all of the derivations given thus far, we have assumed that the pores of the rock are filled with a single-component, single-phase fluid. Oil reservoirs are typically filled with at least two components, oil and water, and often also contain some hydrocarbons in the gaseous phase. We will now present the governing flow equations for an oil-water system, in a fairly general form.

Darcy's law can be generalised for two-phase flow by including a relative permeability factor for each phase:

$$q_w = \frac{-kk_{rw}}{\mu_w}\frac{\partial P_w}{\partial x}, \tag{1.8.1}$$

$$q_o = \frac{-kk_{ro}}{\mu_o}\frac{\partial P_o}{\partial x}, \tag{1.8.2}$$

where the subscripts o and w denote oil and water, respectively. The two relative permeability functions k_{rw} and k_{ro} are assumed to be known functions of the phase saturations. These functions are discussed in detail in the Rock Properties module of this course.

For an oil–water system, the two saturations are necessarily related to each other by

$$S_w + S_o = 1. \tag{1.8.3}$$

In general, the pressures in the two phases at each "point" in the reservoir will be different. If the reservoir is oil-wet, the two pressures will be related by

$$P_o - P_w = P_{\text{cap}}(S_o), \tag{1.8.4}$$

where the capillary pressure P_{cap} is given by some rock-dependent function of the oil saturation. Again, please see the Rock Properties module of this MSc course (Zimmerman, 2017a) for further discussion of the capillary pressure.

As the volume of, say, oil, in a given region is equal to the total pore volume multiplied by the oil saturation, the conservation of mass equations for the two phases can be taken directly from Eq. (1.5.8), by inserting a saturation factor in the storage term:

$$-\frac{\partial(\rho_o q_o)}{\partial x} = \frac{\partial(\phi \rho_o S_o)}{\partial t}, \tag{1.8.5}$$

$$-\frac{\partial(\rho_w q_w)}{\partial x} = \frac{\partial(\phi \rho_w S_w)}{\partial t}. \tag{1.8.6}$$

The densities of the two phases are related to their respective phase pressures by an equation of state:

$$\rho_o = \rho_o(P_o), \tag{1.8.7}$$

$$\rho_w = \rho_w(P_w), \tag{1.8.8}$$

where the right-hand sides of Eqs. (1.8.7) and (1.8.8) are known functions of the pressure, and, for our present purposes, the temperature is assumed constant.

Finally, the porosity must be some function of the two pressures, P_o and P_w. Although the manner in which these two pressures independently affect the porosity is an important and active area of research for soils, which are highly compressible, in oil and gas reservoirs the capillary pressure is always much less than P_o or P_w, and so we can say that $P_o \approx P_w$, in which case we can use the pressure–porosity relationship that would be obtained in a laboratory test performed under single-phase conditions, i.e.

$$\phi = \phi(P_o). \tag{1.8.9}$$

Equations (1.8.1)–(1.8.9) give us nine equations for the nine unknowns (count them!). In many situations, the equations are simplified to allow solutions to be obtained. For example, in the Buckley–Leverett problem of immiscible displacement (discussed in Volume 2 of this series, Blunt, 2017), the densities are assumed to be constant, and the capillary pressure is assumed to be zero.

If the fluid is slightly compressible, or if the pressure variations are small, the equations of state for oil is written as

$$\rho(P_o) = \rho_{oi}[1 + c_o(P_o - P_{oi})], \tag{1.8.10}$$

and similarly for water, where the subscript i denotes the initial state, and the compressibility c_o is taken to be a constant.

Problems for Chapter 1

Problem 1.1. A well located in a 100 ft. thick reservoir having a permeability of 100 mD produces 100 bbl/day of oil from a 10 in. diameter wellbore. The viscosity of the oil is 0.4 cP. The pressure at a distance of 1000 ft. from the wellbore is 3000 psi. What is the pressure at the wellbore? Conversion factors are as follows:

$$\begin{aligned} &1 \text{ barrel} = 0.1589\,\text{m}^3, \\ &1 \text{ Poise} = 0.1 \text{ N-s/m}^2, \\ &1 \text{ foot} = 0.3048\,\text{m}, \\ &1 \text{ psi} = 6895 \text{ N/m}^2 = 6895 \text{ Pa}. \end{aligned}$$

Problem 1.2. Carry out a derivation of the diffusion equation for *spherically-symmetric* flow, in analogy to the derivation given in Section 1.7 for radial flow. (This equation can be used to model flow to a well in situations when only a small length of the well has been perforated, in which case the large-scale flow field will, at early times, be roughly spherical.) The result of your derivation should be an equation similar to Eq. (1.7.8), but with a slightly different term on the right-hand side.

Chapter 2

Line Source Solution for a Vertical Well in an Infinite Reservoir

One of the most basic and important problems in petroleum reservoir engineering, and the cornerstone of well test analysis, is to calculate the pressures in the reservoir, and at the well, when fluid is flowing into a vertical well at a constant rate, from an homogeneous, laterally infinite reservoir. To simplify the problem mathematically, the well is assumed to have an infinitely small radius, i.e. the well is essentially represented by a vertical *line*. The mathematical solution to this important problem will be derived in this chapter.

If fluid were injected into the reservoir from the well, this "line" would serve as a source of fluid for the reservoir; hence, the solution to this problem is referred to the "line source solution". Although the resulting solution for the pressure is more frequently used, by an appropriate change of signs, for the case when fluid is produced *from* the reservoir, in which case the well actually serves as a *sink* as far as the reservoir is concerned, rather than a *source* of fluid, the term "line source" solution is typically used in both cases.

2.1. Derivation of the Line Source Solution

The problem of the flow of a single-phase, slightly compressible fluid to a vertical well in a laterally infinite, homogeneous reservoir can be formulated precisely, as follows:

Geometry: A *vertical* well that *fully penetrates* a reservoir which is of uniform thickness, H, and which extends *infinitely* far in all horizontal directions.

Reservoir Properties: The reservoir is assumed to be *isotropic* and *homogeneous* with uniform properties (i.e. permeability, porosity, etc.) that *do not vary with pressure.*

Initial and Boundary Conditions: The reservoir is initially at a *uniform pressure.* Starting at $t = 0$, fluid is produced from the wellbore at a *constant rate*, Q.

Wellbore Diameter: The diameter of the wellbore is assumed to be *infinitely small*; this leads to a much simpler problem than the more realistic finite-diameter case, but with little loss of applicability, as we will see below.

Problem: To determine the *pressure* at all points in the reservoir, including at the wellbore, as a function of the *elapsed time* since the start of production.

The governing differential equation for this problem is the pressure diffusion equation in radial coordinates, Eq. (1.7.8):

$$\frac{\phi\mu c_t}{k}\frac{\partial P}{\partial t} = \frac{1}{R}\frac{\partial}{\partial R}\left(R\frac{\partial P}{\partial R}\right). \tag{2.1.1}$$

The assumptions that are inherent in this equation are:

(1) The reservoir is homogeneous and isotropic, i.e. k, ϕ, etc., do not vary with position in the reservoir, and so these parameters can be assumed to be constant.
(2) The thickness of the reservoir is uniform; this implies that the flow to the well will be horizontal, with no vertical component.
(3) The well, and the casing perforations, fully penetrate the entire thickness of the reservoir; if not, there would be a vertical flow component.
(4) The fluid is only slightly compressible; this is implicit in treating the compressibility term $c_t = c_f + c_\phi$ as a constant.

If the reservoir is *anisotropic*, the equations can be put into the form of Eq. (2.1.1) by a change of variables that stretches the x and y coordinates (see de Marsily (1986), pp. 178–179).

Inhomogeneity (i.e. spatial variation in k or ϕ) causes great mathematical complications, and methods to treat inhomogeneous reservoirs are still being developed. Variation of the thickness of the reservoir is somewhat equivalent to having a spatial variation in k.

If the well is not perforated over the entire depth of the reservoir, then there would be a vertical flow component. To account for this situation, we would need to add the term $\partial^2 P/\partial z^2$ to the right-hand side of Eq. (2.1.1). The solution to this much more difficult problem is discussed by de Marsily (1986, pp. 179–190).

The case of a highly compressible fluid, such as a gas, whose compressibility varies with pressure, will be discussed in detail in Chapter 9.

To solve the line source problem, or to solve any partial differential equation, we not only need a governing equation, but we also need *initial conditions* and *boundary conditions*. In the present case, these subsidiary conditions are as follows:

Initial Condition: At the start of production, the pressure in the reservoir is assumed to be at some uniform value, P_i.

Boundary Condition at Infinity: Infinitely far from the well, the pressure will always remain at its initial value, P_i.

Boundary Condition at the Wellbore: At the wellbore, which is assumed to be infinitely small, the flow rate must be equal to Q at all times $t > 0$, defined here so that $Q > 0$ for the case of production of fluid *from* the reservoir.

We can therefore formulate the problem in precise mathematical terms as follows, where for notational simplicity we replace c_t with c:

$$\text{Governing PDE:} \quad \frac{1}{R}\frac{\partial}{\partial R}\left(R\frac{\partial P}{\partial R}\right) = \frac{\phi\mu c}{k}\frac{\partial P}{\partial t}, \tag{2.1.2}$$

$$\text{Initial condition:} \quad P(R, t = 0) = P_i, \tag{2.1.3}$$

$$\text{BC at wellbore:} \quad \lim_{R\to 0}\left(\frac{2\pi kH}{\mu}R\frac{\partial P}{\partial R}\right) = Q, \tag{2.1.4}$$

$$\text{BC at infinity:} \quad \lim_{R\to\infty} P(R, t) = P_i. \tag{2.1.5}$$

Strictly speaking, we cannot impose the boundary condition *at* $R = 0$, since at $R = 0$ the term R inside the parenthesis in Eq. (2.1.4) goes to zero, and the term $\partial P/\partial R$ actually goes to infinity. Hence, we need to first multiply these two terms together, and *then* take the "limit" as $R \to 0$.

There are many ways to solve this equation, but we will solve it using a method that does not require advanced techniques such as Laplace transforms, Green's functions, etc. First, we define a new variable η that combines, in a clever way, the spatial variable R and the time variable t. This "trick" for simplifying a diffusion equation was discovered by the German physicist Ludwig Boltzmann in 1894, and this transformation is now referred to as the *Boltzmann transformation*:

$$\eta = \frac{\phi\mu c R^2}{kt}. \tag{2.1.6}$$

We now assume that P will be a function of this *single variable*, η.

Next, we rewrite Eq. (2.1.2) in terms of η. The left-hand side transforms as follows:

$$\frac{\partial P}{\partial R} = \frac{dP}{d\eta}\frac{\partial \eta}{\partial R} = \frac{2\phi\mu c R}{kt}\frac{dP}{d\eta} = \frac{\phi\mu c R^2}{kt}\frac{2}{R}\frac{dP}{d\eta} = \frac{2\eta}{R}\frac{dP}{d\eta}. \tag{2.1.7}$$

Note that since P is a function of the single variable η, the derivative $dP/d\eta$ is an ordinary derivative, not a partial derivative. We therefore see from Eq. (2.1.7) that differentiation with respect to R is equivalent to differentiation with respect to η, followed by multiplication by $2\eta/R$. Hence,

$$\frac{1}{R}\frac{\partial}{\partial R}\left(R\frac{\partial P}{\partial R}\right) = \frac{1}{R}\frac{2\eta}{R}\frac{d}{d\eta}\left(2\eta\frac{dP}{d\eta}\right) = \frac{4\eta}{R^2}\frac{d}{d\eta}\left(\eta\frac{dP}{d\eta}\right). \tag{2.1.8}$$

The right-hand side of Eq. (2.1.2) transforms as follows:

$$\frac{\partial P}{\partial t} = \frac{dP}{d\eta}\frac{\partial \eta}{\partial t} = -\frac{\phi\mu c R^2}{kt^2}\frac{dP}{d\eta} = \frac{-\eta}{t}\frac{dP}{d\eta},$$

$$\to \frac{\phi\mu c}{k}\frac{\partial P}{\partial t} = \frac{-\phi\mu c}{k}\frac{\eta}{t}\frac{dP}{d\eta} = \frac{-\phi\mu c R^2}{kt}\frac{\eta}{R^2}\frac{dP}{d\eta} = \frac{-\eta^2}{R^2}\frac{dP}{d\eta}. \tag{2.1.9}$$

Using Eqs. (2.1.8) and (2.1.9) in Eq. (2.1.2) yields

$$\frac{d}{d\eta}\left(\eta\frac{dP}{d\eta}\right) = -\frac{\eta}{4}\frac{dP}{d\eta}. \tag{2.1.10}$$

Since it contains only one independent variable, η, rather than the two independent variables R and t, Eq. (2.1.10) is an *ordinary differential equation* (ODE) for P as a function of η.

We must now also transform the boundary/initial conditions, so that they apply to the function $P(\eta)$. First, note that both limits, $R \to \infty$ and $t \to 0$, correspond to the same limit, $\eta \to \infty$. Hence, conditions (2.1.3) and (2.1.5) take the form

$$\lim_{\eta \to \infty} P(\eta) = P_i. \tag{2.1.11}$$

Using Eq. (2.1.7) in Eq. (2.1.4) leads to a second BC:

$$\lim_{\eta \to 0} \left(\frac{4\pi k H}{\mu} \eta \frac{dP}{d\eta} \right) = Q$$
$$\to \lim_{\eta \to 0} \left(\eta \frac{dP}{d\eta} \right) = \frac{\mu Q}{4\pi k H}. \tag{2.1.12}$$

The problem is now a *two-point ODE boundary-value problem*, defined by Eqs. (2.1.10–2.1.12).

To solve this problem, we first note that although Eq. (2.1.10) appears to be a second-order differential equation for $P(\eta)$, it is actually a *first-order* equation for the function $\eta\,(dP/d\eta)$. If we temporarily denote $\eta\,(dP/d\eta)$ by the new variable y, we can write Eq. (2.1.10) as

$$\frac{dy}{d\eta} = -\frac{y}{4}, \tag{2.1.13}$$

where

$$y = \eta \frac{dP}{d\eta}.$$

Now separate the variables, and integrate from $\eta = 0$ out to an arbitrary value of η:

$$\frac{dy}{y} = -\frac{d\eta}{4}$$
$$\to \int_{y(0)}^{y(\eta)} \frac{dy}{y} = -\int_0^{\eta} \frac{d\eta}{4}$$

$$\rightarrow \ln\left[\frac{y(\eta)}{y(0)}\right] = -\frac{\eta}{4}$$

$$\rightarrow y(\eta) = y(0)e^{-\eta/4}. \tag{2.1.14}$$

Now note that the boundary condition (2.1.12) is equivalent to

$$y(0) = \frac{\mu Q}{4\pi kH}, \tag{2.1.15}$$

which implies that Eq. (2.1.14) can be written as

$$y(\eta) = \frac{\mu Q}{4\pi kH}e^{-\eta/4}. \tag{2.1.16}$$

Now recall that $y = \eta(dP/d\eta)$, and rewrite Eq. (2.1.16) as

$$\frac{dP(\eta)}{d\eta} = \frac{\mu Q}{4\pi kH}\frac{e^{-\eta/4}}{\eta}. \tag{2.1.17}$$

Equation (2.1.17) can now be directly integrated to find $P(\eta)$, which will give us the pressure in the reservoir as a function of η, and therefore as a function of R and t. We cannot start the integral at $\eta = 0$, because we do not know the pressure at the wellbore. We do, however, know from Eq. (2.1.11) that the pressure at $\eta = \infty$ must be equal to the initial reservoir pressure, P_i. Therefore, we start the integral at $\eta = \infty$:

$$\int_{P_i}^{P(\eta)} dP = \int_{\infty}^{\eta} \frac{\mu Q}{4\pi kH}\frac{e^{-\eta/4}}{\eta}d\eta$$

$$\rightarrow P(\eta) = P_i - \frac{\mu Q}{4\pi kH}\int_{\eta}^{\infty}\frac{e^{-\eta/4}}{\eta}d\eta. \tag{2.1.18}$$

Now recall that $\eta = \phi\mu cR^2/kt$. We replace η with $\phi\mu cR^2/kt$ on the left-hand side of Eq. (2.1.18), and also at the lower limit of integration on the right, but not *inside* the integral because inside the integral η is merely a *dummy integration variable*:

$$P\left(\frac{\phi\mu cR^2}{kt}\right) = P_i - \frac{\mu Q}{4\pi kH}\int_{\frac{\phi\mu cR^2}{kt}}^{\infty}\frac{e^{-\eta/4}}{\eta}d\eta. \tag{2.1.19}$$

Simplify the integrand by defining $u = \eta/4$, in which case $d\eta/\eta = du/u$, and the lower limit of integration becomes $u = \phi\mu c R^2/4kt$:

$$P\left(\frac{\phi\mu c R^2}{4kt}\right) = P_i - \frac{\mu Q}{4\pi kH}\int_{\frac{\phi\mu c R^2}{4kt}}^{\infty} \frac{e^{-u}}{u}du. \tag{2.1.20}$$

The integral in Eq. (2.1.20) is the "exponential integral function", which is defined as (see *Pressure Buildup and Flow Tests in Wells*, Matthews and Russell (1967), p. 131)

$$-Ei(-x) = \int_{x}^{\infty} \frac{e^{-u}}{u}du. \tag{2.1.21}$$

Unfortunately, this function was defined by mathematicians long before it was first used to solve the problem of a well in an infinite reservoir, and so it contains two extraneous minus signs that are awkward, but are now traditional.

Equations (2.1.20) and (2.1.21) give us the pressure as a function of distance from the well and the elapsed time since the start of production. This solution was first presented by the American hydrologist Charles Theis (1935), albeit using a different mathematical approach, and so the line source solution is also frequently referred to as the *Theis solution.*

We can summarise the solution to this problem as follows:

$$P(R,t) = P_i + \frac{\mu Q}{4\pi kH}Ei(-x), \tag{2.1.22}$$

where

$$-Ei(-x) = \int_{x}^{\infty} \frac{e^{-u}}{u}du, \tag{2.1.23}$$

and

$$x = \phi\mu c R^2/4kt. \tag{2.1.24}$$

Numerical values for the pressure at some location in the reservoir are found as follows. Assume that we want to know the pressure at a certain distance R from the centre of the well, at some time t. We use these values of R and t to compute x from Eq. (2.1.24). We then look up the value of $-Ei(-x)$ from a table or graph of the exponential integral function (see below). The pressure at (R, t) is then given by Eq. (2.1.22).

The more common situation is that the pressure is measured at some distance from the well, or at the well itself, as a function of time, and the pressure data are used to infer the values for the reservoir parameters, by fitting the data to the analytical solution. This procedure will be demonstrated later, after we first analyse the line source solution in more detail.

Note: To find the pressure at the wellbore, we merely plug $R = R_w$ into Eqs. (2.1.22–2.1.24)! This is because $R = R_w$ corresponds to the point in the reservoir located just at wellbore wall, where the pressure must be the same as the pressure of the fluid in the wellbore.

Numerical values of the Ei function are shown in Table 2.1, taken from *Quantitative Hydrogeology* by de Marsily (1986).

For example, if $x = 5 \times 10^{-7}$, then $-Ei(-x) = 13.93$.

2.2. Dimensionless Pressure and Time

Although the pressure seems to depend on many variables and parameters, there are actually only two *independent, dimensionless* mathematical variables in the line source solution. This can be proven from the pi-theorem of dimensional analysis, or can be seen directly from Eqs. (2.1.22–2.1.24).

Traditionally, these variables are defined as the *dimensionless time*

$$t_D = \frac{kt}{\phi \mu c R^2}, \tag{2.2.1}$$

and the *dimensionless pressure drawdown*,

$$\Delta P_D = \frac{2\pi k H (P_i - P)}{\mu Q}. \tag{2.2.2}$$

In terms of these dimensionless parameters, the line source solution takes the form (see Figure 2.1)

$$\Delta P_D = -\frac{1}{2} Ei(-1/4t_D). \tag{2.2.3}$$

Table 2.1. Exponential integral function, $-Ei(-x)$.

x	1	2	3	4	5	6	7	8	9
$\times 1$	0.219	0.049	0.013	0.0038	0.0011	$3.6e{-4}$	$1.2e{-4}$	$3.8e{-5}$	$1.2e{-5}$
$\times 10^{-1}$	1.82	1.22	0.91	0.70	0.56	0.45	0.37	0.31	0.26
$\times 10^{-2}$	4.04	3.35	2.96	2.68	2.47	2.30	2.15	2.03	1.92
$\times 10^{-3}$	6.33	5.64	5.23	4.95	4.73	4.54	4.39	4.26	4.14
$\times 10^{-4}$	8.63	7.94	7.53	7.25	7.02	6.84	6.69	6.55	6.44
$\times 10^{-5}$	10.94	10.24	9.84	9.55	9.33	9.14	8.99	8.86	8.74
$\times 10^{-6}$	13.24	12.55	12.14	11.85	11.63	11.45	11.29	11.16	11.04
$\times 10^{-7}$	15.54	14.85	14.44	14.15	13.93	13.75	13.60	13.46	13.34
$\times 10^{-8}$	17.84	17.15	16.74	16.46	16.23	16.05	15.90	15.76	15.65
$\times 10^{-9}$	20.15	19.45	19.05	18.76	18.54	18.35	18.20	18.07	17.95
$\times 10^{-10}$	22.45	21.76	21.35	21.06	20.84	20.66	20.50	20.37	20.25
$\times 10^{-11}$	24.75	24.06	23.65	23.36	23.14	22.96	22.81	22.67	22.55
$\times 10^{-12}$	27.05	26.36	25.96	25.67	25.44	25.26	25.11	24.97	24.86
$\times 10^{-13}$	29.36	28.66	28.26	27.97	27.75	27.56	27.41	27.28	27.16
$\times 10^{-14}$	31.66	30.97	30.56	30.27	30.05	29.87	29.71	29.58	29.46
$\times 10^{-15}$	33.96	33.27	32.86	32.58	32.35	32.17	32.02	31.88	31.76

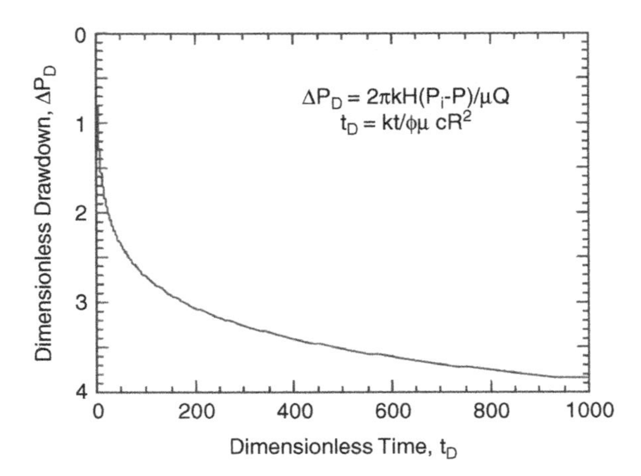

Figure 2.1. Line source solution plotted in terms of dimensionless time and dimensionless drawdown.

Recall that the line source solution allows us to calculate the pressure at any radius R. Hence, by the above definitions, the dimensionless time is different at each location R in the reservoir. Most often, however, we are interested in the pressure at the well, in which case $R = R_w$, and the dimensionless time at the well is given by $t_{\mathrm{Dw}} = kt/\phi\mu c R_w^2$.

The usefulness of dimensionless variables is that they allow the pressure drawdown to be plotted and discussed in a form that is applicable to all reservoirs, without being restricted to specific values of the permeability, porosity, etc. These latter parameters are accounted for by the definitions of the dimensionless variables.

Note: It follows from Eq. (1.6.12) that, aside from a factor of 4, the dimensionless time at a distance R from the centre of the borehole is equal to the actual physical time t, normalised against the time required for the peak of the pressure pulse to travel from the borehole out to radial location R. Hence, values of $4t_D \ll 1$ correspond to times at which the pressure pulse has not yet arrived at location R, whereas times such that $4t_D \gg 1$ correspond to times at which the pressure pulse has penetrated much farther than a distance R into the reservoir.

2.3. Range of Applicability of the Line Source Solution

The line source solution assumes that the wellbore radius is "zero", when in reality it is of course non-zero. Does this cause a problem in practice? Fortunately, the answer is: no, not really! We can use the line source solution as soon as the "radius of penetration" of the pressure pulse, as predicted by the line source solution, is greater than R_w, the actual wellbore radius. This seems reasonable and can be proven rigorously by examining the solution to the diffusion equation with finite wellbore radius (which will be presented in Chapter 6).

According to Eq. (1.6.12), the time required for the pressure pulse to travel at least a distance R_w, starting from the hypothetically "infinitely-small" borehole at $R = 0$, is

$$t > \frac{\phi \mu c R_w^2}{4k}. \tag{2.3.1}$$

If we use "typical" values for the parameters, such as $\phi = 0.2$ (a typical reservoir value), $\mu = 0.001\,\text{Pa s}$ (order of magnitude for liquid hydrocarbons), $c = 10^{-10}\,\text{Pa}^{-1}$ (reasonable value for liquid hydrocarbons), $R_w = 0.1\,\text{m}$ (order of magnitude of typical borehole), $k = 10^{-14}\,\text{m}^2$ (10 mD; somewhat low, but possible), then Eq. (2.3.1) predicts that the line source approximation will become valid after an elapsed time of only 0.005 s!

In terms of the dimensionless time defined by Eq. (2.2.1), the condition given above by Eq. (2.3.1) is equivalent to $t_{\text{Dw}} > 0.25$, where the subscript w denotes the fact that this is the dimensionless time relative to the wellbore, where $R = R_w$.

Hence, the validity of the line source approximation is not compromised by the mathematical assumption of an "infinitely small wellbore radius". However, other physical effects, such as wellbore storage, cause the line source solution to be inaccurate at small times; these effects are discussed in Chapter 5.

2.4. Logarithmic Approximation to the Line Source Solution

The exponential integral function is unfamiliar to most engineers and is difficult to calculate. Fortunately, for sufficiently small values of x, which is to say, large values of t, the exponential integral essentially becomes a *logarithmic function*, which makes it very easy to use.

To derive this "late-time" approximation, we proceed as follows. For large times, x will be small, and we can break up the integral into two parts:

$$-Ei(-x) = \int_x^\infty \frac{e^{-u}}{u} du = \int_x^1 \frac{e^{-u}}{u} du + \int_1^\infty \frac{e^{-u}}{u} du. \qquad (2.4.1)$$

Use the Taylor series for $\exp(-u)$ in the first integrand on the right:

$$\int_x^1 \frac{e^{-u}}{u} du = \int_x^1 \frac{1 - \frac{u}{1!} + \frac{u^2}{2!} - \frac{u^3}{3!} + \cdots}{u} du. \qquad (2.4.2)$$

Break up the integral on the right side of Eq. (2.4.2) into a series of integrals, and evaluate them term-by-term:

$$\begin{aligned}
&\int_x^1 \frac{e^{-u}}{u} du \\
&= \int_x^1 \frac{1}{u} du - \frac{1}{1!} \int_x^1 du + \frac{1}{2!} \int_x^1 u du + \frac{1}{3!} \int_x^1 u^2 du - \cdots \\
&= \ln u]_x^1 - u]_x^1 + \frac{1}{2!} \frac{u^2}{2}\bigg]_x^1 - \frac{1}{3!} \frac{u^3}{3}\bigg]_x^1 + \cdots \\
&= (\ln 1 - \ln x) - (1 - x) + \frac{1}{2!2}(1 - x^2) - \frac{1}{3!3}(1 - x^3) + \cdots \\
&= -\ln x + x - \frac{1}{2!2}x^2 + \frac{1}{3!3}x^3 + \cdots - \left\{1 - \frac{1}{2!2} + \frac{1}{3!3} + \cdots\right\}.
\end{aligned} \qquad (2.4.3)$$

Substituting this result this back into Eq. (2.4.1) allows us to say

$$-Ei(-x) = -\ln x - \ln \gamma + x - \frac{1}{2!2}x^2 + \frac{1}{3!3}x^3 + \cdots, \qquad (2.4.4)$$

where

$$\ln \gamma = \left\{1 - \frac{1}{2!2} + \frac{1}{3!3} + \cdots\right\} - \int_1^\infty \frac{e^{-u}}{u} du.$$

Equation (2.4.4) looks messy, but the important point is that $\ln \gamma$ is merely a number, and does *not depend* on x, so we can evaluate it numerically, once and for all, to find

$$\ln \gamma = \ln(1.781) = 0.5772. \tag{2.4.5}$$

Note: Both γ *and* $\ln \gamma$ are sometimes known as "Euler's number" in the petroleum engineering literature, so care must be taken when reading other papers and books. Although most mathematicians use γ to denote the number 0.5772, in these notes, γ will always denote 1.781.

We can further simplify Eq. (2.4.4) if we can find conditions under which the power series terms are negligible. If this is the case, we will be left with only a *logarithmic* term and a *constant.* First, recall that large t implies small x, and so

$$x - \frac{1}{2!2}x^2 + \frac{1}{3!3}x^3 + \cdots < x. \tag{2.4.6}$$

If we want the power series terms to be, say, two orders of magnitude less than γ, which itself is roughly on the order of 1, then we need

$$x = \frac{\phi\mu c R^2}{4kt} < 0.01 \rightarrow \frac{kt}{\phi\mu c R^2} > 25. \tag{2.4.7}$$

So, if the dimensionless time is greater than about 25, we have, from Eqs. (2.1.22) and (2.4.4):

$$P(R,t) = P_i + \frac{\mu Q}{4\pi k H}(\ln x + \ln \gamma)$$

$$\rightarrow P(R,t) = P_i + \frac{\mu Q}{4\pi k H}\ln(x\gamma)$$

$$\rightarrow P(R,t) = P_i + \frac{\mu Q}{4\pi k H}\ln\left(\frac{\phi\mu c R^2 \gamma}{4kt}\right)$$

$$\rightarrow P(R,t) = P_i - \frac{\mu Q}{4\pi kH} \ln\left(\frac{4kt}{\phi\mu cR^2\gamma}\right)$$

$$\rightarrow P(R,t) = P_i - \frac{\mu Q}{4\pi kH} \ln\left(\frac{2.246kt}{\phi\mu cR^2}\right) \qquad (2.4.8)$$

$$\rightarrow P(R,t) = P_i - \frac{\mu Q}{4\pi kH} \left[\ln\left(\frac{kt}{\phi\mu cR^2}\right) + 0.80907\right]. \qquad (2.4.9)$$

The form given in Eq. (2.4.8) is used in groundwater hydrology, and is called *Jacob's approximation*; the equivalent form of Eq. (2.4.9) is used in petroleum reservoir engineering, where it is called the *logarithmic approximation.*

By comparing Eq. (2.4.9) with Eqs. (2.2.1–2.2.3), we see that the dimensionless form of the logarithmic approximation is

$$\Delta P_D = \frac{1}{2}\left[\ln(t_D) + 0.80907\right]. \qquad (2.4.10)$$

For values of the dimensionless time that are typically of interest at the well, the logarithmic approximation is very accurate. It can be seen in Figure 2.2 that the range of validity of the logarithmic approximation is consistent with the criterion given in Eq. (2.4.7),

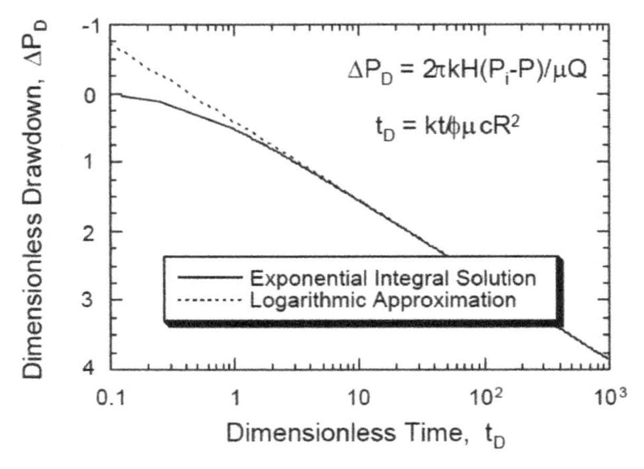

Figure 2.2. Comparison of the logarithmic approximation to the full exponential integral solution.

i.e. there is excellent agreement for $t_D > 25$. Although smaller values of t_D are usually not of interest at the well, they *are* needed if one wants to calculate the drawdown at a location far from the well. For values of $t_D < 1$, the logarithmic approximation is actually quite inaccurate. Please verify this for yourself for, say, $t_D = 0.25$, by evaluating Eq. (2.4.9), and comparing the result to that obtained from Table 2.1.

2.5. Instantaneous Pulse of Injected Fluid

Insight can be gained into the *diffusive* nature of flow through a porous medium by considering the problem of a finite amount of fluid injected into a well over a very small period of time. This problem also introduces the important concept of superposition, which will be described more fully in Chapter 3.

Note: The problem of "injecting" fluid is mathematically equivalent to that of "producing" fluid, except for the sign, but it is probably easier to visualise a positive pressure pulse propagating into the reservoir, as occurs during injection, rather than a "negative" pressure pulse propagating into the reservoir during production.

If we start *injecting* fluid at a rate Q (m^3/s) at time $t = 0$, then, according to Eq. (2.1.22), the pressure at a distance R into the reservoir will be

$$P(R,t) = P_i + \frac{\mu Q}{4\pi kH}\int_x^\infty \frac{e^{-u}}{u}du, \quad x = \phi\mu cR^2/4kt, \tag{2.5.1}$$

where t is the *elapsed time since the start of injection*, and Q is defined here to be a positive number. The + sign appears in Eq. (2.5.1) because, if we are *injecting* rather than *extracting* fluid, the pressure in the reservoir should be *greater* than the initial reservoir pressure.

Now imagine that we *stop* injecting fluid after a small amount of time, δt. This is equivalent to *injecting* fluid at a rate Q starting at $t = 0$, and then *producing* fluid at a rate Q (or, equivalently, injecting at a rate $-Q$) starting at time δt. The pressure drawdown in the reservoir due to this fictitious production would be given by the same line source solution, *except that:*

(1) For extraction of fluid we must use a "–" sign in front of the integral;
(2) If t is the elapsed time since the start of injection, then $t - \delta t$ will be the elapsed time since the start of the fictitious extraction of fluid (i.e. since the end of the actual injection!)

Hence, the full expression for the pressure at location R and time t will be

$$P(R,t) = P_i + \frac{\mu Q}{4\pi kH}\int_{x=\frac{\phi\mu cR^2}{4kt}}^{\infty}\frac{e^{-u}}{u}du - \frac{\mu Q}{4\pi kH}\int_{x=\frac{\phi\mu cR^2}{4k(t-\delta t)}}^{\infty}\frac{e^{-u}}{u}du$$

$$= P_i + \frac{\mu Q}{4\pi kH}\int_{x=\frac{\phi\mu cR^2}{4kt}}^{x=\frac{\phi\mu cR^2}{4k(t-\delta t)}}\frac{e^{-u}}{u}du. \quad (2.5.2)$$

But if δt is small, then the two limits of integration in the integral on the right-hand side of Eq. (2.5.2) are close together, and we can use the following approximation:

$$\int_{x_1}^{x_2} f(x)dx \approx f(x_1)[x_2 - x_1], \quad (2.5.3)$$

which in the present case gives

$$P(R,t) \approx P_i + \frac{\mu Q}{4\pi kH}\cdot\frac{4kt}{\phi\mu cR^2}\cdot e^{-\frac{\phi\mu cR^2}{4kt}}\cdot\left[\frac{\phi\mu cR^2}{4k(t-\delta t)} - \frac{\phi\mu cR^2}{4kt}\right]$$

$$\approx P_i + \frac{\mu Q}{4\pi kH}\cdot\frac{4kt}{\phi\mu cR^2}\cdot e^{-\frac{\phi\mu cR^2}{4kt}}\cdot\left[\frac{\phi\mu cR^2\delta t}{4kt^2}\right]$$

$$\approx P_i + \frac{\mu Q\delta t}{4\pi kHt}e^{-\frac{\phi\mu cR^2}{4kt}}. \quad (2.5.4)$$

As Q is the rate of injection in (m^3/s), and δt is the duration of the injection, the total volume of injected fluid is $Q\delta t$, which we can hereafter denote as $Q*$, with units of (m^3).

Imagine now that we are monitoring the pressure at some fixed distance R from the borehole, for example, with a pressure gauge in an observation well. Equation (2.5.4) shows that the pressure buildup

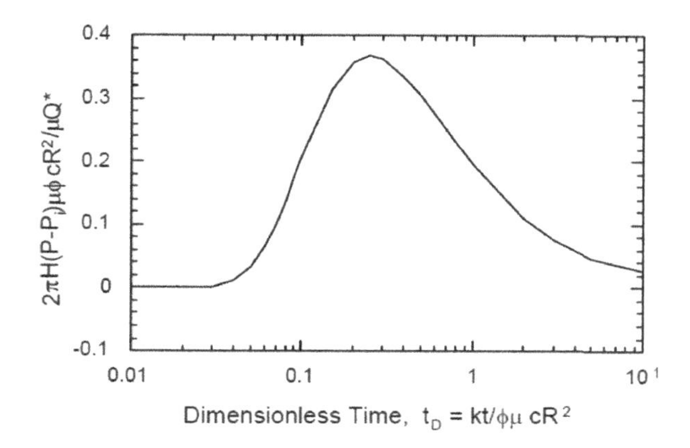

Figure 2.3. Pressure buildup due to an injected pulse of fluid.

at R will be the product of *two* terms: an exponential term that increases with t, and then levels off to the value of 1 as $t \to \infty$; and a term proportional to $1/t$, that decays to zero as $t \to \infty$. The product of these increasing and decreasing terms creates a function that first increases with time, and then decreases, as shown in Figure 2.3.

It seems reasonable to identify the "time at which the pressure pulse has arrived at location R" with the time at which the pressure buildup reaches its *maximum* value at R. This time can be found by setting $\partial P/\partial t = 0$ in Eq. (2.5.4):

$$\left.\frac{\partial P}{\partial t}\right|_R = \frac{\mu Q^*}{4\pi kH}\left[\frac{-1}{t^2} + \frac{\phi\mu cR^2}{4kt^3}\right] e^{-\frac{\phi\mu cR^2}{4kt}}, \tag{2.5.5}$$

$$\left.\frac{\partial P}{\partial t}\right|_R = 0 \Rightarrow t = \frac{\phi\mu cR^2}{4k}, \tag{2.5.6}$$

which provides a derivation of Eq. (1.6.12).

2.6. Estimating Permeability and Storativity from a Drawdown Test

In Sections 2.1–2.4, we derived the line source solution, and showed how to use it to calculate the pressure drawdown, based on assumed knowledge of the reservoir properties. However, the most common use of this solution, and the other solutions that we will derive in subsequent chapters, is for the *inverse* problem:

- We use measured wellbore pressures, in conjunction with the mathematical solutions, to estimate reservoir properties such as permeability, porosity, etc.
- This process, known as *well test analysis*, is the subject of a subsequent module of this MSc course. For now, we will do one simple example to see how to calculate the reservoir permeability from a drawdown test.

Recall from Eq. (2.4.8) that, in the late-time regime,

$$P(R,t) = P_i - \frac{\mu Q}{4\pi kH} \ln\left(\frac{2.246kt}{\phi\mu c R^2}\right). \tag{2.6.1}$$

At the wellbore wall,

$$P(R_w,t) = P_i - \frac{\mu Q}{4\pi kH} \left[\ln t + \ln\left(\frac{2.246k}{\phi\mu c R_w^2}\right)\right]. \tag{2.6.2}$$

In general, we will not know the values of most of the parameters on the right-hand side of Eq. (2.6.2); we will only know Q, and the wellbore pressure as a function of time, $P(R_w,t) \equiv P_w(t)$. In particular, we do not know k or $\phi\mu c$, so we cannot use criterion (2.4.7) to find out when our data fall into the late-time regime.

However, the second logarithmic term, although unknown, is a *constant*. Hence, if we plot $P_w(t)$ versus $\ln t$, the data will eventually, at large enough values of t, fall on a straight line! The slope of this line on a semi-log plot gives kH, i.e.

$$\left|\frac{dP_w}{d\ln t}\right| = \left|\frac{\Delta P_w}{\Delta \ln t}\right| \equiv m = \frac{\mu Q}{4\pi kH}, \tag{2.6.3}$$

$$\rightarrow kH = \frac{\mu Q}{4\pi m}. \tag{2.6.4}$$

In practice, Q is known, and μ can be measured, so the semi-log slope m gives us the permeability-thickness product, kH.

Note that this method is unable to distinguish separately between the effects of *permeability* and *thickness*; i.e. a thick reservoir of low permeability can give the same drawdown as a more permeable but thinner reservoir.

Table 2.2. Drawdown data used to illustrate the method for determining permeability.

t (mins)	1	5	10	20	30	60
P_w (psi)	4740	4667	4633	4596	4573	4535

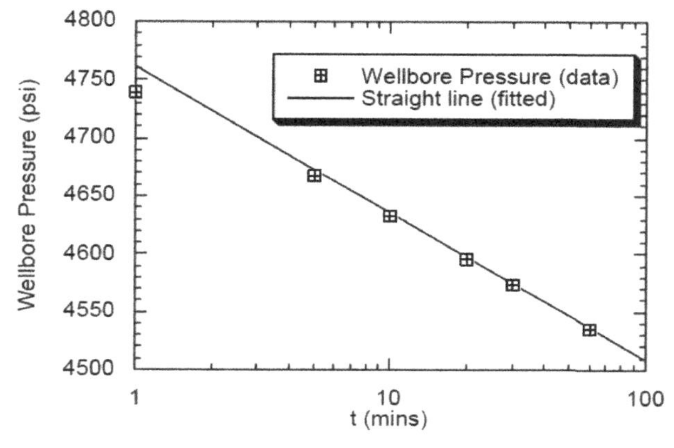

Figure 2.4. Semi-log straight line method for estimating the permeability from a drawdown test.

Now, let us do a simple example to learn how to calculate the permeability of a reservoir from a drawdown test.

Example: A well with 4 in. radius produces oil with viscosity 0.3 cP, at a constant rate of 200 bbl/day, from a reservoir that is 15 ft. thick. The wellbore pressure as a function of time is given in Table 2.2. Use the "semi-log straight line" method to estimate the permeability, k.

(1) Plot wellbore pressure against the logarithm of time, as in Figure 2.4.
(2) Look for a straight line at *late* times, and find its slope:

$$m = \left| \frac{\Delta P}{\Delta \ln t} \right| = \frac{4760 - 4510}{2 \times 2.303} = 54.3\,\text{psi}$$

$$= 54.3\,\text{psi} \times \frac{6895\text{Pa}}{\text{psi}} = 374,400\ \text{Pa}.$$

Note: $\Delta \ln t$ will have the same value, *regardless* of which units are used for t!

(3) Calculate k from Eq. (2.6.4), first converting all data to SI:

$$\mu = 0.3 \text{ cP} \times \frac{0.001 \text{ Pa s}}{\text{cP}} = 0.0003 \, \text{Pa s},$$

$$Q = 200 \frac{\text{bbl}}{\text{day}} \times \frac{0.1589 \, \text{m}^3}{\text{bbl}} \times \frac{\text{day}}{24 \, \text{hr}} \times \frac{\text{hr}}{3600 \, \text{s}} = 3.68 \times 10^{-4} \frac{\text{m}^3}{\text{s}},$$

$$H = 15\text{ft} \times \frac{0.3048 \, \text{m}}{\text{ft}} = 4.572 \, \text{m}$$

$$\rightarrow k = \frac{\mu Q}{4\pi m H} = \frac{(0.0003 \, \text{Pa s})(3.68 \times 10^{-4} \text{m}^3/\text{s})}{4\pi (4.572 \, \text{m})(374,400 \, \text{Pa})}$$

$$= 5.13 \times 10^{-15} \text{m}^2 \times \frac{1\text{mD}}{0.987 \times 10^{-15} \text{m}^2} = 5.1 \, \text{mD}.$$

We can also use a semi-log plot to estimate the storativity term, ϕc. To see how this works, first note that on a semi-log plot of P_w versus $\ln t$, there are *two* asymptotic straight lines (see Figure 2.2). At early times, $P_w = P_i$ (i.e. a horizontal line). At late times, P_w slopes *downward* as a function of $\ln t$, according to Eq. (2.6.3).

At what time t^* do these straight lines intersect? This occurs when P_w(early t asymptote) $= P_w$(late t asymptote), i.e.

$$P_i = P_i - \frac{\mu Q}{4\pi k H} \ln \left(\frac{2.246 k t*}{\phi \mu c R_w^2} \right)$$

$$\rightarrow \ln \left(\frac{2.246 k t*}{\phi \mu c R_w^2} \right) = 0$$

$$\rightarrow \frac{2.246 k t*}{\phi \mu c R_w^2} = 1$$

$$\rightarrow t* = \frac{\phi \mu c R_w^2}{2.246 k},$$

which can be inverted to give the storativity

$$\phi c = \frac{2.246 k t*}{\mu R_w^2}. \tag{2.6.5}$$

Problems for Chapter 2

Problem 2.1. A well with 3 in. radius is located in a 40 ft. thick reservoir that has a permeability of 30 mD and a porosity of 0.20. The total compressibility of the oil-rock system is 3×10^{-5}/psi. The initial pressure in the reservoir is 2800 psi. The well produces 448 bbl/day of oil that has a viscosity of 0.4 cP. Conversion factors can be found in Problem 1.1.

(a) How long will it take in order for the line source solution to be applicable at the wellbore wall?
(b) What is the pressure at the wellbore after six days of production, according to the line source solution?
(c) How long will it take in order for Jacob's logarithmic approximation to be valid at the wellbore?
(d) What is the pressure at the wellbore after six days of production, according to the logarithmic approximation?
(e) Answer questions (b)–(d) for a location that is 800 ft. (horizontally) away from the wellbore.

Problem 2.2. A well with a radius of 0.3 ft. produces 200 bbl/day of oil, with viscosity 0.6 cP, from a 20 ft. thick reservoir. The wellbore pressures are given in the table below. Estimate the permeability and the storativity of the reservoir, using the semi-log method described in Section 2.6.

t(mins)	0	5	10	20	60	120	480	1440	2880	5760
P_w(psi)	4000	3943	3938	3933	3926	3921	3911	3904	3899	3894

Chapter 3

Superposition and Pressure Buildup Tests

During well tests, the well may be flowed for a period of time, and then "shut-in". It may also be flowed at a sequence of different rates, to obtain data that can be used to determine various reservoir properties. These sequences result in a complex pressure signal. In this chapter, we will introduce the principle of superposition, and show how it can be used to develop methods for analysing these complex pressure signals, and thereby help us to infer the values of important reservoir properties such as permeability and storativity.

3.1. Linearity and the Principle of Superposition

A basic property of the pressure diffusion equation that governs flow of a single-phase compressible liquid through a porous medium is its *linearity*. Linearity is the most important and useful property that any differential equation can have because it allows the principle of *superposition* to be used to construct solutions to the equation. Most of the analytical methods that have been developed to solve differential equations, such as Laplace transforms, Green's functions, eigenfunction expansions, etc., can in fact be used *only* on linear differential equations. These analytical methods will be discussed to some extent in later sections of these notes. In this section, we will discuss a simple form of the principle of superposition that will allow us to solve many important reservoir engineering problems, such as pressure buildup tests.

The diffusion equation is a *linear* partial differential equation (PDE) because both of the differential operators that appear in it are *linear operators*. Formally, a differential operator M that operates on a function F is linear if it has the following two properties:

$$M(F_1 + F_2) = M(F_1) + M(F_2), \tag{3.1.1}$$

$$M(cF_1) = cM(F_1), \tag{3.1.2}$$

where F_1 and F_2 are *any* two differentiable functions, and c is any constant.

The process of partial differentiation is a linear operation, since

$$\frac{d}{dt}[P_1(R,t) + P_2(R,t)] = \frac{dP_1(R,t)}{dt} + \frac{dP_2(R,t)}{dt}, \tag{3.1.3}$$

$$\frac{d}{dt}[cP_1(R,t)] = c\frac{dP_1(R,t)}{dt}. \tag{3.1.4}$$

By this definition, we see that Eq. (1.7.8) is a linear PDE. This will be the case if the coefficients that appear in the diffusivity equation, such as ϕ, c, μ and k, are constants. If these coefficients varied with position or time, the governing equation (see Section 1.8) would still be linear, albeit more difficult to solve.

However, if any of the coefficients were functions of *pressure*, the equation would no longer be linear. This situation occurs, for example, with gas flow, for which the compressibility varies with pressure (see Chapter 9). It is also the case for "stress-sensitive" reservoirs in which the permeability varies with pressure.

A simple rule-of-thumb is that a differential equation will be nonlinear if it contains any term in which the dependent variable (in our case, P) or any of its derivatives appear to a power higher than one, or are multiplied by one another. For example, $M = P(dP/dt)$ is a nonlinear operator, because it violates condition (3.1.1):

$$\begin{aligned} M\{P_1 + P_2\} &= (P_1 + P_2)\frac{d(P_1 + P_2)}{dt} \\ &= (P_1 + P_2)\left[\frac{dP_1}{dt} + \frac{dP_2}{dt}\right] \end{aligned}$$

$$= P_1 \frac{dP_1}{dt} + P_2 \frac{dP_1}{dt} + P_1 \frac{dP_2}{dt} + P_2 \frac{dP_2}{dt}$$

$$= M\{P_1\} + M\{P_2\} + P_2 \frac{dP_1}{dt} + P_1 \frac{dP_2}{dt}$$

$$\rightarrow M\{P_1 + P_2\} \neq M\{P_1\} + M\{P_2\}! \tag{3.1.5}$$

The importance of linearity is that it allows us to create new solutions to the diffusion equation by adding together previously known solutions. Care must be taken, however, with the initial conditions and boundary conditions. For example, if P_1 and P_2 are two pressure functions that each satisfy the diffusion equation (2.1.2) and the initial condition (2.1.3), then the sum of P_1 and P_2 will also satisfy the diffusion equation, but will *not* satisfy the correct initial conditions, because

$$P_1(R, t = 0) + P_2(R, t = 0) = P_i + P_i = 2P_i. \tag{3.1.6}$$

This difficulty can be circumvented by working with the *drawdown* instead of the actual pressure. Linearity of the diffusion equation implies that the drawdown, $\Delta P = P_i - P(R, t)$, satisfies the same diffusion equation as does $P(R, t)$ itself, since, for example,

$$\frac{d[P_i - P(R,t)]}{dt} = \frac{dP_i}{dt} - \frac{dP(R,t)}{dt} = -\frac{dP(R,t)}{dt}, \quad \text{etc.} \tag{3.1.7}$$

As the drawdown satisfies zero initial condition, by definition, Eq. (3.1.6) shows that the sum of two drawdown functions will also satisfy the correct initial condition (i.e. that the drawdown must be zero when $t = 0$). Likewise, the drawdown is also zero infinitely far from the well, so if two drawdown functions satisfy the boundary condition at $R = \infty$, their sum will also satisfy this far-field boundary condition.

3.2. Pressure Buildup Test in an Infinite Reservoir

In a *pressure buildup test*, a well that has been producing fluid at a constant rate Q for some time t is then "shut-in", i.e. production is stopped. After this, fluid will continue to flow towards the well, due to the pressure gradient in the reservoir, but will not be able to exit at the wellhead. Consequently, the pressure at the well will rise back towards it initial value, P_i. The rate of this pressure recovery at the well can be used to estimate both the transmissivity, kH, and the initial pressure, P_i, of the reservoir.

The analysis of a pressure buildup test is based on the *principle of superposition* that was discussed in the previous section, and proceeds as follows. First, imagine that we produce at a rate Q, starting at $t = 0$, in which case the pressure drawdown *due to this production* will be

$$\Delta P_1 = P_i - P_1(R,t) = -\frac{\mu Q}{4\pi kH} Ei\left(\frac{-\phi\mu c R^2}{4kt}\right). \tag{3.2.1}$$

Now consider the following fictitious problem, in which, at some time t_1, we begin to *inject* fluid into the same well, at rate Q. The pressure drawdown due to this injection would be given by the same line source solution, *except* that:

(a) The variable that we use in the line source solution to represent the "elapsed time" must be measured from the start of injection, i.e. the variable must be $t - t_1$;
(b) Since we are injecting rather than producing, we must use "$-Q$" in the solution.

Therefore, the pressure drawdown due to this fictitious injection is

$$\Delta P_2 = P_i - P_2(R,t) = \frac{\mu Q}{4\pi kH} Ei\left[\frac{-\phi\mu c R^2}{4k(t - t_1)}\right]. \tag{3.2.2}$$

Note: It is implicitly understood in all equations such as Eq. (3.2.2) that the value of the Ei function is taken to be zero when the term in brackets is positive, which is to say, when $t < t_1$.

We now superimpose these two solutions (for the drawdown, not the pressure itself!), putting $\Delta P = \Delta P_1 + \Delta P_2$. In light of the discussion given in the previous section, this composite function is also a solution to the pressure diffusion equation. However, we note that:

(a) For $t < t_1$, the injection has not started, and so the composite solution corresponds to production at rate Q, as given by Eq. (3.2.1).
(b) For $t < t_1$, the composite solution corresponds to production at rate Q, and injection at rate Q, which is to say, a well that is neither producing nor injecting, i.e. it is "shut-in"!
(c) Therefore, this superposition of the two pressure functions solves the problem of production for a time t_1, followed by shut-in:

$$\begin{aligned} \Delta P(R,t) &= \Delta P_1(R,t) + \Delta P_2(R,t) \\ &= -\frac{\mu Q}{4\pi kH} Ei\left(\frac{-\phi\mu cR^2}{4kt}\right) + \frac{\mu Q}{4\pi kH} Ei\left[\frac{-\phi\mu cR^2}{4k(t-t_1)}\right] \\ &= -\frac{\mu Q}{4\pi kH}\left\{Ei\left(\frac{-\phi\mu cR^2}{4kt}\right) - Ei\left[\frac{-\phi\mu cR^2}{4k(t-t_1)}\right]\right\}. \end{aligned} \tag{3.2.3}$$

Now recall that $\Delta P(R,t) = P_i - P(R,t)$, in which case $P(R,t) = P_i - \Delta P(R,t)$, and so

$$P(R,t) = P_i + \frac{\mu Q}{4\pi kH}\left\{Ei\left(\frac{-\phi\mu cR^2}{4kt}\right) - Ei\left[\frac{-\phi\mu cR^2}{4k(t-t_1)}\right]\right\}. \tag{3.2.4}$$

If t is sufficiently large that we can use the logarithmic approximation for *both* terms in Eq. (3.2.4), then

$$\begin{aligned} P(R,t) &= P_i - \frac{\mu Q}{4\pi kH}\left[\ln t + \ln\left(\frac{2.246k}{\phi\mu cR^2}\right)\right. \\ &\qquad \left. - \ln(t-t_1) - \ln\left(\frac{2.246k}{\phi\mu cR^2}\right)\right] \\ &\rightarrow P(R,t) = P_i - \frac{\mu Q}{4\pi kH}[\ln t - \ln(t-t_1)] \\ &\rightarrow P(R,t) = P_i - \frac{\mu Q}{4\pi kH}\ln\frac{t}{(t-t_1)}. \end{aligned} \tag{3.2.5}$$

This is the equation for the pressure at the wellbore during a pressure buildup test. However, the notation more commonly used in buildup tests is the following:

- The duration of production period is denoted by t, not t_1.
- The duration of the shut-in period is denoted by Δt, not $t - t_1$.

Using this notation, the well pressure during shut-in is given by

$$P_w(t) = P_i - \frac{\mu Q}{4\pi kH} \ln\left(\frac{t + \Delta t}{\Delta t}\right). \tag{3.2.6}$$

This equation is used for graphical analysis of the data from a buildup test, allowing estimation of the kH product and the initial reservoir pressure. The ratio $(t + \Delta t)/\Delta t$ that appears in Eq. (3.2.6) is known as the *Horner time*, t_H. Note that the Horner time has the following peculiar properties:

(a) It does not have units of time, but is *dimensionless*, and
(b) It becomes numerically *smaller* as the duration of the shut-in period increases.

The wellbore pressure that would be measured during a shut-in test is shown schematically in Figure 3.1.

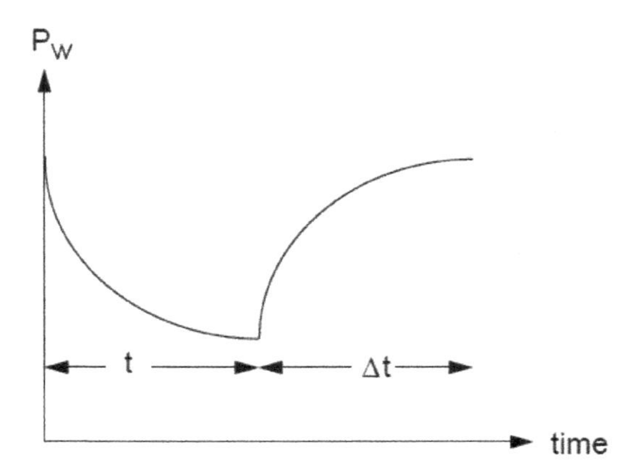

Figure 3.1. Wellbore pressure during a buildup test.

3.3. Multi-rate Flow Tests

The superposition principle that was used to solve the problem of a buildup test can also be used in the more general situation in which the production rate is changed by discrete amounts at various time intervals. First, imagine that the production rate is given by

$$Q = Q_0 \quad \text{for } 0 < t < t_1, \tag{3.3.1}$$

$$Q = Q_1 \quad \text{for } t > t_1. \tag{3.3.2}$$

To find the drawdown, we superpose the solution for production at rate Q_0 starting at time $t = 0$, plus a solution starting at t_1 that corresponds to the *increment* in the production rate, $Q_1 - Q_0$:

$$\Delta P(R,t) = -\frac{\mu Q_0}{4\pi kH} Ei\left(\frac{-\phi\mu cR^2}{4kt}\right) - \frac{\mu(Q_1 - Q_0)}{4\pi kH} Ei\left[\frac{-\phi\mu cR^2}{4k(t-t_1)}\right]. \tag{3.3.3}$$

To verify that it is correct to use the flow rate *increment*, note that for $t > t_1$, the first Ei function corresponds to a flow rate of Q_0, and the second corresponds to a rate of $Q_1 - Q_0$, so the total flow rate is

$$Q(t > t_1) = Q_0 + (Q_1 - Q_0) = Q_1. \tag{3.3.4}$$

The drawdown in the general case in which flow rate Q_i commences at time t_i can therefore be represented by

$$\Delta P(R,t) = \frac{-\mu Q_0}{4\pi kH} Ei\left(\frac{-\phi\mu cR^2}{4kt}\right) - \sum_{i=1} \frac{\mu(Q_i - Q_{i-1})}{4\pi kH} Ei\left[\frac{-\phi\mu cR^2}{4k(t-t_i)}\right]. \tag{3.3.5}$$

We can simplify the notation by defining the pressure drawdown *per unit of flow rate*, with production starting at $t = 0$, as $\Delta P_Q(R,t)$. For the line source in an infinite reservoir, this definition takes the form

$$\Delta P_Q(R,t) \equiv \frac{\Delta P(R,t;Q)}{Q} \equiv \frac{-\mu}{4\pi kH} Ei\left(\frac{-\phi\mu c R^2}{4kt}\right), \tag{3.3.6}$$

with the understanding that $\Delta P_Q(R,t) = 0$ when $t < 0$. Using definition (3.3.6), the drawdown in a multi-rate test can be written as

$$\Delta P(R,t) = Q_0 \Delta P_Q(R,t) + \sum_{i=1} (Q_i - Q_{i-1}) \Delta P_Q(R, t - t_i). \tag{3.3.7}$$

This type of multi-rate test is used, for example, to determine rate-dependent skin factors (see Chapter 5).

3.4. Convolution Integral for Variable-rate Flow Tests

The superposition formula (3.3.7) can be generalised further, to the case where the flow rate at the well is some arbitrary (but continuous) function of time, $Q(t)$. We first note that an arbitrary production schedule can always be approximated by a discrete number of time periods during which the flow rate is constant, as shown in Figure 3.2.

Now recall that the time derivative of the flow rate, at time t_i, where i is an index that takes on the values 1, 2, 3, etc., can be

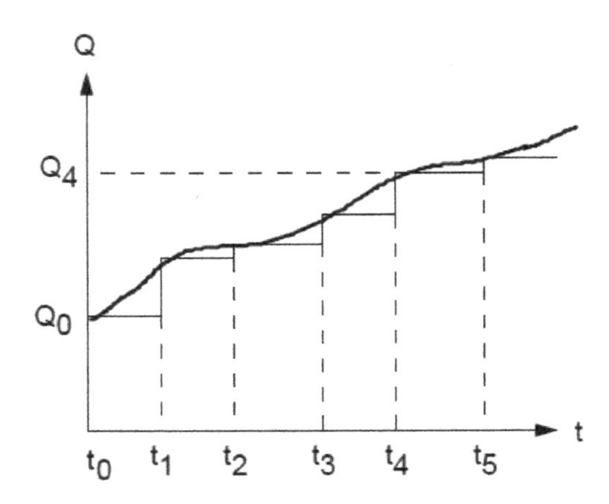

Figure 3.2. Variable-rate production approximated by multi-rate production.

approximated as

$$\left.\frac{dQ}{dt}\right|_{t_i} \approx \frac{(\Delta Q)_i}{(\Delta t)_i} = \frac{Q_i - Q_{i-1}}{t_i - t_{i-1}}. \tag{3.4.1}$$

Equation (3.4.1) can be rearranged, to allow us to approximate the flow rate increment, as follows:

$$Q_i - Q_{i-1} \approx \frac{dQ(t_i)}{dt} \times (t_i - t_{i-1}). \tag{3.4.2}$$

Using this approximation in Eq. (3.3.7) gives

$$\Delta P(R,t) = Q_0 \Delta P_Q(R,t) + \sum_{i=1} \frac{dQ(t_i)}{dt} \Delta P_Q(R, t - t_i)(t_i - t_{i-1}). \tag{3.4.3}$$

We now simplify the notation by rewriting Eq. (3.4.3) in the following equivalent form:

$$\Delta P(R,t) = Q_0 \Delta P_Q(R,t) + \sum_{i=1} Q'(t_i) \Delta P_Q(R, t - t_i) \Delta t_i. \tag{3.4.4}$$

As we make each time increment smaller, approximation (3.4.1) becomes more accurate, and the "step-function" approximation to the flow rate $Q(t)$ also becomes more accurate. In the limit as each time increment goes to zero, the errors due to these approximations will vanish.

Furthermore, as the time increments get smaller, the series in Eq. (3.4.4) becomes an *integral* with respect to t_i:

$$\Delta P(R,t) = Q_0 \Delta P_Q(R,t) + \int_{t_i=0}^{t_i=t} Q'(t_i) \Delta P_Q(R, t - t_i) dt_i. \tag{3.4.5}$$

The integral in Eq. (3.4.5) ends at $t_i = t$ because when $t_i > t$, the function $\Delta P_Q(R, t - t_i)$ is zero, by definition. Physically, this is equivalent to the fact that a change in flow rate that occurs at a time *later than* t cannot possibly have an effect on the drawdown *at* time t.

Finally, we note that in the limit as each of the time increments become infinitely small, the finite number of times that we were denoting by t_i evolve into a continuous variable, which we will denote by τ. The drawdown can then be written as

$$\Delta P(R,t) = Q_0 \Delta P_Q(R,t) + \int_0^t \frac{dQ(\tau)}{d\tau} \Delta P_Q(R,t-\tau) d\tau. \qquad (3.4.6)$$

The integral in Eq. (3.4.6) is known as a *convolution integral*, and is specifically referred to as "the convolution of the two functions dQ/dt and ΔP_Q". Formula (3.4.6) is also known as Duhamel's principle, after the French mathematician who first suggested this procedure in the context of solving heat conduction problems (Duhamel, 1833).

The importance of the convolution integral given by Eq. (3.4.6) is that it allows us to find the drawdown for any production schedule, by merely performing a single integral utilising the "constant flow rate" solution. For example, if we have a reservoir with, say, a closed outer circular boundary (see Chapter 6), then we need only to find the solution for the case of constant flow rate in a reservoir with a closed outer circular boundary; the solution for a variable flow rate in this reservoir then follows from Eq. (3.4.6), with the constant flow rate solution playing the role of the function ΔP_Q.

Another form of the convolution integral that is sometimes more convenient to use can be derived by applying integration-by-parts to the integral in (3.4.6). First, recall the general expression for integration by parts:

$$\int_0^t f(\tau) \frac{dg(\tau)}{d\tau} d\tau = f(\tau) g(\tau)]_0^t - \int_0^t g(\tau) \frac{df(\tau)}{d\tau} d\tau. \qquad (3.4.7)$$

We now put $f(\tau) = \Delta P_Q(R, t-\tau)$ and $g(\tau) = Q(\tau)$, in which case Eq. (3.4.6) becomes

$$\Delta P(R,t) = Q_0 \Delta P_Q(R,t) + Q(\tau) \Delta P_Q(R,t-\tau)]_0^t \\ - \int_0^t Q(\tau) \frac{d\Delta P_Q(R,t-\tau)}{d\tau} d\tau,$$

$$= Q_0 \Delta P_Q(R,t) + Q(t)\Delta P_Q(R,0) - Q(0)\Delta P_Q(R,t)$$
$$- \int_0^t Q(\tau)\frac{d\Delta P_Q(R,t-\tau)}{d\tau}d\tau. \tag{3.4.8}$$

But $Q(0) = Q_0$, by definition, so the first and third terms on the right cancel out. And $\Delta P_Q(R,0)$ is the drawdown at time zero, which must be zero, and so the second term on the right also drops out.

Next, use of the chain rule shows that

$$\frac{d\Delta P_Q(R,t-\tau)}{d\tau} = \frac{-d\Delta P_Q(R,t-\tau)}{dt}, \tag{3.4.9}$$

in which case Eq. (3.4.8) can finally be written as

$$\Delta P(R,t) = \int_0^t Q(\tau)\frac{d\Delta P_Q(R,t-\tau)}{dt}d\tau. \tag{3.4.10}$$

This integral allows us to calculate the drawdown for *any* production history, provided that we know the drawdown function for the case of *constant production rate*, $\Delta P_Q(R,t)$. It implies that, for a given reservoir shape, we only ever need to solve the problem of constant production rate!

Problems for Chapter 3

Problem 3.1. Which, if any, of the following differential equations are linear, and why (or why not)?

(a) $\frac{d^2y}{dx^2} + y\frac{dy}{dx} + y = 0.$

(b) $\frac{d^2y}{dx^2} + x\frac{dy}{dx} + y = 0.$

(c) $\frac{d^2y}{dx^2} + x\frac{dy}{dx} + xy = 0.$

Problem 3.2. Find an expression for the wellbore pressure in a vertical well in a laterally infinite reservoir, if the production rate increases linearly as a function of time according to $Q(t) = Q_* t/t_*$, where Q_* and t_* are constants. Use convolution, in the form of either Eq. (3.4.6) or Eq. (3.4.10), and recall that $\Delta P_Q(R,t)$ for a well in an infinite reservoir is given by Eq. (3.3.6).

Chapter 4

Effect of Faults and Linear Boundaries

In the previous two chapters, we assumed that the well was located in a laterally infinite reservoir. But all real reservoirs are bounded. For example, many reservoirs contain nearly vertical faults which, due to mineralisation processes, or to the creation of fault gouge caused by relative motion between the two sides of the fault, may serve as impermeable barriers to flow. In this chapter, we will learn how to use spatial superposition of fictitious "image" wells to model the effect that such impermeable faults will have on the pressure behaviour measured at the actual production well. In particular, this analysis will allow us to calculate the distance from the well to the nearest impermeable sealing fault. The method of image wells can also be used to model the effect of linear constant-pressure boundaries, such as may be caused by an aquifer abutting against the reservoir.

4.1. Superposition of Sources/Sinks in Space

In Section 3.1, we saw that if we add together two different line source solutions that "start" at different times, we still have a legitimate mathematical solution to the pressure diffusion equation. We can also add together solutions that represent line sources that are located at different places in *space*, and their sum will represent a solution to the diffusion equation.

The simplest and most straightforward example of the use of spatial superposition is for the problem of two wells located in an infinite reservoir, as shown in Figure 4.1. Well 1 is located at point

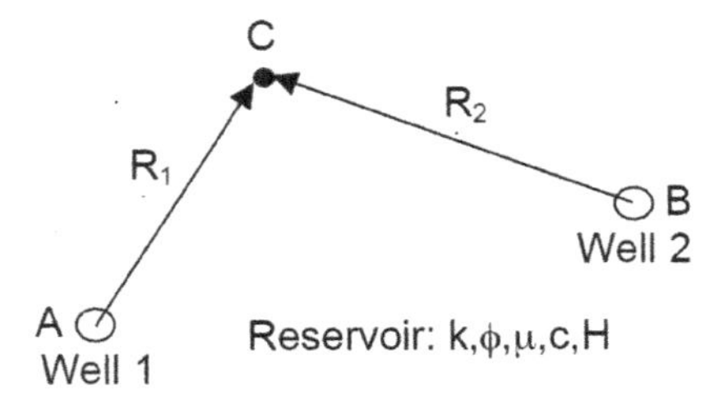

Figure 4.1. Two wells producing from an infinite, homogeneous reservoir, with the pressure monitored at an observation well C.

A, and the distance from A to a generic point C in the reservoir is denoted by R_1. Well 2 is located at point B, and the distance from B to C is denoted by R_2. Well 1 starts producing at rate Q_1 at time t_1, and well 2 starts producing at rate Q_2, starting at time t_2. It may be convenient to imagine an observation well located at C, fitted with a pressure gauge.

We now claim that the drawdown at arbitrary point C in the reservoir, at some time t, is given by the sum of the two relevant line source solutions

$$P(R,t) = P_i + \frac{\mu Q_1}{4\pi kH} Ei\left[\frac{-\phi\mu c R_1^2}{4k(t-t_1)}\right] + \frac{\mu Q_2}{4\pi kH} Ei\left[\frac{-\phi\mu c R_2^2}{4k(t-t_2)}\right]. \tag{4.1.1}$$

To prove that this is indeed the proper solution to this problem, we note that:

(1) The line source solution for well 1 satisfies the diffusion equation at all points in the reservoir (except at $R_1 = 0$, where it gives an "infinite" drawdown), and at all times. Likewise for the second line source solution. Hence, their sum satisfies the pressure diffusion equation at all times, and at all points except $R_1 = 0$ and $R_2 = 0$. But these two points are not actually located in the reservoir (they are located "in the well"), so we do not care that the diffusion equation is not satisfied there!

(2) Both line source solutions satisfy the initial condition of zero drawdown, so their sum also satisfies this initial condition.

(3) Both line source solutions satisfy the far-field boundary condition that the drawdown is zero "at infinity", and so their sum

also satisfies this boundary condition, i.e.

$$P(R \to \infty, t) = P_i + \frac{\mu Q}{4\pi kH} Ei(-\infty) + \frac{\mu Q}{4\pi kH} Ei(-\infty) = P_i, \tag{4.1.2}$$

because, as can be seen from Table 2.1, $Ei(-x)$ goes to zero as x becomes very large. Hence, we have verified that the drawdown given by Eq. (4.1.1) is the correct solution for the two-well problem. Obviously, spatial superposition can be used for any number of wells.

The principle of superposition can also be used in conjunction with the concept of "image" wells, to solve problems such as finding the effect of a nearby impermeable fault, as explained in the next section.

4.2. Effect of an Impermeable Vertical Fault

Hydrocarbon reservoirs are often transected by faults, many of which are nearly vertical. Due to mineral deposition on the fault surfaces, accumulation of fault gouge and other geological processes, faults are often *impermeable* to flow. In order to properly interpret the results of well tests, it is important to understand the effect that an impermeable boundary will have on a drawdown test. It is therefore desirable to find the solution to the problem of constant-rate production from a well located in a homogeneous reservoir that is bounded on one side by a vertical fault of infinite extent, and which is impermeable to flow. This solution can easily be found using the Theis solution for a well in an infinite *unbounded* reservoir, and using the principle of spatial superposition.

Consider a well located at a perpendicular distance (i.e. nearest distance) d from an impermeable fault, which appears in planview as a straight line that extends infinitely far in both directions, as in Figure 4.2. This well produces fluid at a constant rate Q, starting at $t = 0$. Now imagine a fictitious "image well" that is situated as the "mirror image" of the first well (i.e. located at a distance d on the other side of the fault), which also produces fluid from the reservoir at rate Q starting at $t = 0$. In this hypothetical situation, we ignore

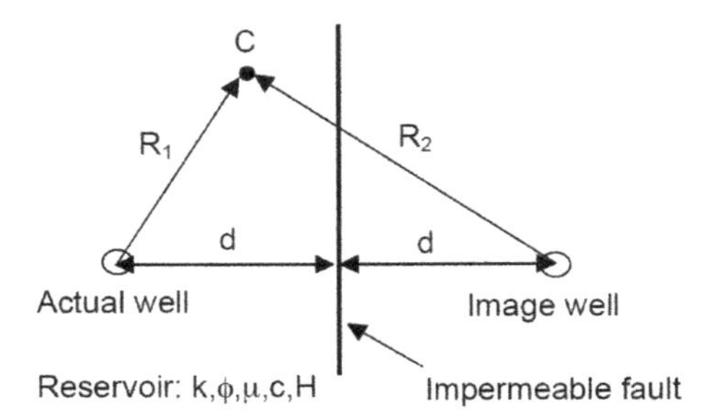

Figure 4.2. Use of an image well to solve the problem of a well near an impermeable fault.

the fact that the fault exists, and assume that the reservoir extends infinitely far in all horizontal directions, just as in the Theis problem. Due to the symmetry of this situation, no fluid will ever cross the plane of the fault. Hence, this plane will effectively act as a *no-flow boundary.* This scenario therefore gives us the solution for a well near an impermeable fault!

The pressure drawdown in this situation is the superposition of the drawdown due to the *actual* well, plus the drawdown due to the fictitious *image* well:

$$P(R,t) = P_i + \frac{\mu Q}{4\pi kH} Ei\left(\frac{-\phi\mu c R_1^2}{4kt}\right) + \frac{\mu Q}{4\pi kH} Ei\left(\frac{-\phi\mu c R_2^2}{4kt}\right). \tag{4.2.1}$$

To find the pressure in the wellbore, we put

$$R_1 = R_w, \quad R_2 = 2d - R_w \approx 2d, \tag{4.2.2}$$

in which case Eq. (4.2.1) becomes

$$P_w(t) = P_i + \frac{\mu Q}{4\pi kH} Ei\left(\frac{-\phi\mu c R_w^2}{4kt}\right) + \frac{\mu Q}{4\pi kH} Ei\left[\frac{-\phi\mu c (2d)^2}{4kt}\right]. \tag{4.2.3}$$

There are several time regimes to consider:

(i) An early time regime, during which the logarithmic approximation is not yet valid for either the actual well solution, or for the

image-well solution. From Eq. (2.4.7), this regime is defined by

$$t < \frac{25\phi\mu c R_w^2}{k}, \quad \text{or} \quad t_{\mathrm{Dw}} < 25. \tag{4.2.4}$$

In this regime, we would have to use the full infinite series expansion for the exponential integral function. However, we saw in Section 2.4 that the duration of this regime is usually very short, and so there is no need for us to study it further.

(ii) An intermediate time regime in which the logarithmic approximation can be used for the well solution, but the drawdown in the wellbore *due to the image solution* is still negligible. The *start* of this regime is defined by Eq. (4.2.4), i.e.

$$t > \frac{25\phi\mu c R_w^2}{k}, \quad \text{or} \quad t_{\mathrm{Dw}} > 25. \tag{4.2.5}$$

This regime ends when the *front edge* of the pressure pulse from the image well reaches the actual well, which is located at a distance $2d$ away from the image well. We can consider that this occurs when the second Ei function in Eq. (4.2.3) reaches, say, 0.01. According to Table 2.1, this will occur when the argument of the Ei function reaches about 3.3. Therefore, the upper limit of this second time regime is defined by

$$t < \frac{0.3\phi\mu c d^2}{k}, \quad \text{or} \quad t_{\mathrm{Dw}} < 0.3(d/R_w)^2. \tag{4.2.6}$$

In this regime, the pressure in the well is given by

$$P_w(t) = P_i - \frac{\mu Q}{4\pi k H} \ln\left(\frac{2.246kt}{\phi\mu c R_w^2}\right). \tag{4.2.7}$$

This is precisely the drawdown that would occur in the absence of the fault, because in this time regime the actual pressure pulse has not yet had time to travel to the fault and reflect back to the wellbore (where it can be detected). In particular, during this regime the slope on a semi-log plot of the wellbore pressure curve will be

$$\frac{dP_w}{d\ln t} = \frac{\Delta P_w}{\Delta \ln t} = \frac{-\mu Q}{4\pi k H}. \tag{4.2.8}$$

(iii) A late time regime during which the logarithmic approximation is valid for *both* the actual well solution, *and* the image well solution. This regime is defined by

$$t > \frac{25\phi\mu c(2d)^2}{k}, \quad \text{or} \quad t_{\mathrm{Dw}} > 100(d/R_w)^2. \tag{4.2.9}$$

In this regime, the pressure at the actual well is given by

$$\begin{aligned} P_w(t) &= P_i - \frac{\mu Q}{4\pi kH}\ln\left(\frac{2.246kt}{\phi\mu c R_w^2}\right) - \frac{\mu Q}{4\pi kH}\ln\left[\frac{2.246kt}{\phi\mu c(2d)^2}\right] \\ &= P_i - \frac{\mu Q}{4\pi kH}\left[\ln t + \ln\left(\frac{2.246k}{\phi\mu c R_w^2}\right) + \ln t + \ln\left(\frac{2.246k}{\phi\mu c4d^2}\right)\right] \\ &= P_i - \frac{\mu Q}{4\pi kH}\left[2\ln t + \ln\left(\frac{2.246k}{\phi\mu c R_w^2}\right) + \ln\left(\frac{2.246k}{\phi\mu c4d^2}\right)\right]. \end{aligned} \tag{4.2.10}$$

This equation will also yield a straight line on a plot of P_w versus $\ln t$, but with a slope that is *twice* that of the earlier slope, i.e.

$$\frac{dP_w}{d\ln t} = \frac{\Delta P_w}{\Delta \ln t} = \frac{-2\mu Q}{4\pi kH} = \frac{-\mu Q}{2\pi kH}. \tag{4.2.11}$$

Doubling of the slope of the graph of P_w versus $\ln t$ is therefore an indication of a nearby impermeable boundary. Moreover, the distance from the production well to the fault can be estimated from the transition time between the regime when Eq. (4.2.7) is applicable, to the regime when Eq. (4.2.10) becomes applicable (see Problem 4.1).

Physical Explanation: Because of the impermeable boundary, oil is being produced from a "semi-infinite" reservoir, rather than an infinite reservoir. So, at large times, only *half* as much oil will be produced, for a given wellbore pressure. Therefore, in order to maintain a constant flow rate, we must *double* the drawdown.

4.3. Two Intersecting Impermeable Vertical Faults

The method of images can also be used to study the effect of two or more intersecting faults. For example, consider a well located

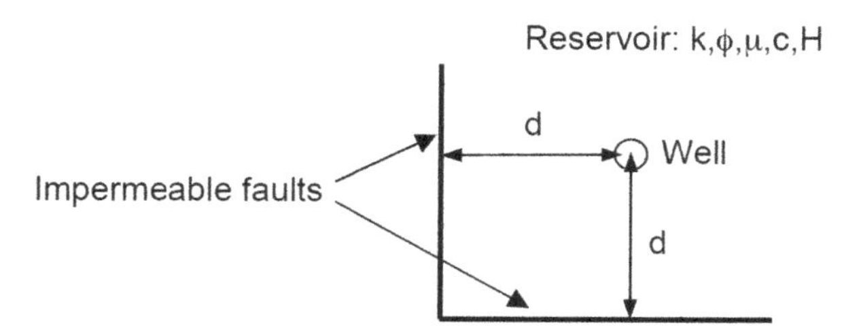

Figure 4.3. Well located near two intersecting impermeable faults.

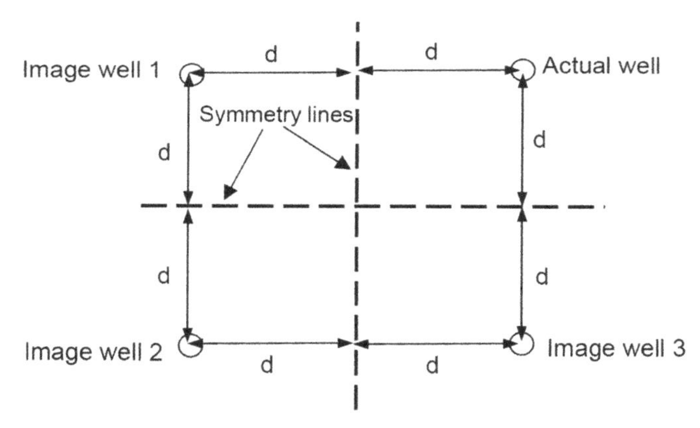

Figure 4.4. Use of image wells to solve problem of a well near two intersecting impermeable faults.

equidistant from two impermeable faults that intersect at a right angle, as in Figure 4.3.

To solve this problem, we again first imagine that the reservoir is laterally infinite in extent, and contains no impermeable boundaries. However, we want the locations of the actual impermeable boundaries to correspond to "no-flow" boundaries in our fictitious unbounded reservoir. This will be the case if we have two perpendicular lines of symmetry located at distances d from the actual well. This can be accomplished by introducing the following *three* images wells, as shown in Figure 4.4.

Image wells 1 and 3 are both located at a distance $2d$ from the actual well, whereas the distance from the actual well to image well 2 is $2\sqrt{2}d$. Hence, we must add *three* "fictitious" line source solutions to the line source solution for the actual well, each with the appropriate radius variable. The resulting drawdown at the actual production well is given by

$$\frac{4\pi kH[P_w(t)-P_i]}{\mu Q} = Ei\left(\frac{-\phi\mu c R_w^2}{4kt}\right) + 2Ei\left(\frac{-\phi\mu c 4d^2}{4kt}\right) + Ei\left(\frac{-\phi\mu c 8d^2}{4kt}\right). \tag{4.3.1}$$

Recalling the definitions of dimensionless time and dimensionless drawdown given in Section 2.2, we can write Eq. (4.3.1) in dimensionless form as

$$\Delta P_{\mathrm{Dw}} = -\frac{1}{2}Ei\left(\frac{-1}{4t_{\mathrm{Dw}}}\right) - Ei\left[\frac{-(d/R_w)^2}{t_{\mathrm{Dw}}}\right] - \frac{1}{2}Ei\left[\frac{-2(d/R_w)^2}{t_{\mathrm{Dw}}}\right]. \tag{4.3.2}$$

The wellbore pressure will exhibit an early stage in which the effect of the image wells is not yet apparent, and the semi-log slope of the wellbore pressure versus time curve will be $-\mu Q/4\pi kH$. Eventually, at a sufficiently long time such that the drawdown in the actual well due to production from *all four wells* can be approximated by the logarithmic approximation, the semi-log slope will be *four times* greater in magnitude, i.e. $-\mu Q/\pi kH$.

Note that application of the method of images is not as simple as saying "if we had one impermeable boundary we added one image well, so for two impermeable boundaries we should add two image wells"; this statement might seem logical, but it is not correct. The trick is to place image wells so as to create a fictitious unbounded reservoir that has the proper symmetry in its flow field.

Several other examples of the use of the method of images, such as two infinite parallel faults, two faults intersecting at 45°, etc., can be found in the book *Well Testing in Heterogeneous Formations*, by Streltsova (1988).

4.4. Well Near a Linear Constant-pressure Vertical Boundary

Another situation that sometimes arises in reservoirs is that of a linear *constant-pressure* boundary. This boundary may, for example, be formed by a gas cap that borders a slightly dipping oil reservoir. The effect that such a boundary has on a drawdown test can also be studied by using the method of images.

In this case, imagine an image well, located as in Figure 4.2, but *injecting* fluid at a rate Q. Again, we "ignore" the actual constant-pressure boundary, and assume that the actual well and the image well are located in an infinite reservoir. By superposition, the drawdown at a generic location in the reservoir will in this case be

$$P(R,t) = P_i + \frac{\mu Q}{4\pi kH} Ei\left(\frac{-\phi\mu c R_1^2}{4kt}\right) - \frac{\mu Q}{4\pi kH} Ei\left(\frac{-\phi\mu c R_2^2}{4kt}\right). \tag{4.4.1}$$

Now consider a point on the mid-plane between the two wells. At such a point, $R_1 = R_2$, and so Eq. (4.4.1) gives

$$\begin{aligned} P(\text{midplane}, t) &= P_i + \frac{\mu Q}{4\pi kH}\left[Ei\left(\frac{-\phi\mu c R_1^2}{4kt}\right) - Ei\left(\frac{-\phi\mu c R_1^2}{4kt}\right)\right] \\ &= P_i \quad \text{for all } t \end{aligned} \tag{4.4.2}$$

which proves that the mid-plane is indeed a plane of *constant pressure.*

Now, consider the pressure drawdown in the actual well, which is found by setting $R_1 = R_w$ and $R_2 = 2d - R_w \approx 2d$

$$P(R,t) = P_i + \frac{\mu Q}{4\pi kH} Ei\left(\frac{-\phi\mu c R_w^2}{4kt}\right) - \frac{\mu Q}{4\pi kH} Ei\left[\frac{-\phi\mu c (2d)^2}{4kt}\right]. \tag{4.4.3}$$

Again, aside from a very-early time regime defined by Eq. (4.2.4), which we ignore, we have, in analogy with Eqs. (4.2.5) and (4.2.6), a regime defined by $t_{\rm Dw} > 25$ and

$$t < \frac{0.3\phi\mu c d^2}{k}, \quad \text{or} \quad t_{\rm Dw} < 0.3(d/R_w)^2, \tag{4.4.4}$$

during which the effect of the image well has not yet been felt in the actual well, and so the pressure in the well is given by

$$P_w(t) = P_i - \frac{\mu Q}{4\pi k H}\ln\left(\frac{2.246kt}{\phi\mu c R_w^2}\right). \tag{4.4.5}$$

There exists a late-time regime during which the logarithmic approximation is valid at the production well for both the actual line source solution and the image line source solution. In analogy with Eq. (4.2.9), this regime is defined by

$$t > \frac{25\phi\mu c(2d)^2}{k}, \quad \text{or} \quad t_{\rm Dw} > 100(d/R_w)^2. \tag{4.4.6}$$

In this regime, the pressure in the production well is given by

$$\begin{aligned}
P_w(t) &= P_i - \frac{\mu Q}{4\pi k H}\ln\left(\frac{2.246kt}{\phi\mu c R_w^2}\right) + \frac{\mu Q}{4\pi k H}\ln\left[\frac{2.246kt}{\phi\mu c(2d)^2}\right] \\
&= P_i - \frac{\mu Q}{4\pi k H}\left\{\ln t + \ln\left(\frac{2.246k}{\phi\mu c R_w^2}\right) - \ln t - \ln\left[\frac{2.246k}{\phi\mu c(2d)^2}\right]\right\} \\
&= P_i - \frac{\mu Q}{4\pi k H}\left\{\ln\left(\frac{2.246k}{\phi\mu c}\right) - \ln(R_w^2)\right. \\
&\quad \left. - \ln\left(\frac{2.246k}{\phi\mu c}\right) + \ln[(2d)^2]\right\} \\
&= P_i - \frac{\mu Q}{4\pi k H}\ln[(2d/R_w)^2] \\
&= P_i - \frac{\mu Q}{2\pi k H}\ln\left(\frac{2d}{R_w}\right).
\end{aligned} \tag{4.4.7}$$

We see that the pressure at the production well eventually stabilises to a constant value. The steady-state value of the drawdown depends on the dimensionless distance to the boundary (i.e. relative to the wellbore radius).

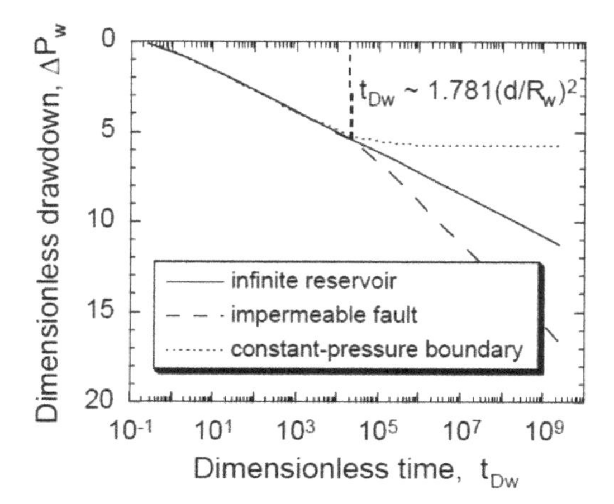

Figure 4.5. Wellbore pressure for a well near a vertical fault.

Recall that the slope of P_w versus $\ln t$ was $-\mu Q/4\pi kH$ for a well in an infinite reservoir. An impermeable linear boundary eventually causes an *additional* slope of $-\mu Q/4\pi kH$, leading to a total late-time slope of $-\mu Q/2\pi kH$. On the other hand, a constant-pressure linear boundary causes and "additional" slope of $+\,\mu Q/4\pi kH$, leading to a late-time slope of 0, as shown in Figure 4.5.

Problems for Chapter 4

Problem 4.1. As explained in Section 4.2, a doubling of the slope on a semi-log plot of drawdown versus time indicates the presence of an impermeable linear fault. The drawdown data can also be used to find the distance from the well to the fault, as follows. If we plot the data and then fit two straight lines through the early-time and late-time data, the time at which these lines intersect is called $t'_{\rm Dw}$. Show that the distance to the fault is then given by the equation $d = (0.5615t'_{\rm Dw})^{1/2}R_w$.

Problem 4.2. The curves in Figure 4.5 were drawn for the case $d = 200R_w$. What would the curves look like for the case of a fault located at a distance $d = 400R_w$?

Problem 4.3. Consider a well located equidistant from two orthogonal boundaries, as in Figure 4.3, but imagine that the boundaries are *constant-pressure* boundaries, rather than impermeable boundaries. How would you utilise the method of images to find the drawdown in this well?

Chapter 5

Wellbore Skin and Wellbore Storage

In the line source solution that we have been discussing thus far, all effects associated with the borehole itself are actually ignored, and the borehole is idealised as an infinitely thin "line". In this chapter, we will examine two physical phenomena associated with the borehole that necessitate a more detailed analysis of flow to a well. One of these phenomena is the existence of a zone of altered permeability around the borehole; the second is the effect of transient fluid storage within the borehole itself.

5.1. Wellbore Skin Concept: Steady-state Model

The solutions to the pressure diffusion equation that were presented in previous chapters were all based on the assumption that the permeability of the reservoir is *uniform* in space. However, it is often the case that the rock immediately surrounding the wellbore has a lower permeability than the remainder of the reservoir, for various reasons:

(1) Infiltration of drilling mud into the formation, which clogs up the pores of the rock.
(2) Swelling of clays in the formation due to contact with drilling fluid.

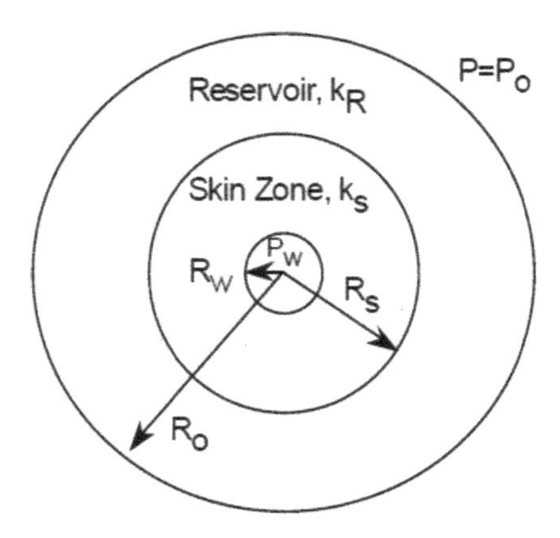

Figure 5.1. "Skin" zone surrounding a wellbore.

(3) Incomplete perforation of the wellbore casing, which causes the flowpaths of the fluid to be constricted as they approach the wellbore. This has an effect that is similar to that caused by a low-permeability zone near the wellbore.

To account for these possibilities, we introduce the concept of "wellbore skin", which can be thought of as a thin annular region surrounding the borehole in which the permeability k_s is lower than that of the undamaged reservoir rock, as illustrated schematically in Figure 5.1. The permeability in this idealised composite reservoir is *discontinuous*, with $k(R)$ given by

$$R_w < R < R_s \rightarrow k = k_s, \tag{5.1.1}$$

$$R_s < R < R_o \rightarrow k = k_R. \tag{5.1.2}$$

To quantify the effect of the wellbore skin, consider the steady-state radial flow model that was discussed in Section 1.4. The governing equation is the radial version of Darcy's law

$$Q = \frac{2\pi kH}{\mu} R \frac{dP}{dR}, \tag{5.1.3}$$

where we use the convention that $Q > 0$ for production.

We can separate the variables of Eq. (5.1.3), and then integrate from $R = R_w$ to $R = R_o$, bearing in mind that k depends on R:

$$\frac{1}{k}\frac{dR}{R} = \frac{2\pi H}{\mu Q}dP$$

$$\int_{R_w}^{R_s}\frac{1}{k_s}\frac{dR}{R} + \int_{R_s}^{R_o}\frac{1}{k_R}\frac{dR}{R} = \int_{P_w}^{P_o}\frac{2\pi H}{\mu Q}dP$$

$$\frac{1}{k_s}\ln\frac{R_s}{R_w} + \frac{1}{k_R}\ln\frac{R_o}{R_s} = \frac{2\pi H}{\mu Q}(P_o - P_w)$$

$$\frac{1}{k_R}\left[\ln\frac{R_o}{R_s} + \frac{k_R}{k_s}\ln\frac{R_s}{R_w}\right] = \frac{2\pi H}{\mu Q}(P_o - P_w)$$

$$\frac{1}{k_R}\left[\ln\frac{R_o}{R_s} + \ln\frac{R_s}{R_w} + \frac{k_R}{k_s}\ln\frac{R_s}{R_w} - \ln\frac{R_s}{R_w}\right] = \frac{2\pi H}{\mu Q}(P_o - P_w)$$

$$\frac{1}{k_R}\left[\ln\left(\frac{R_o}{R_s}\frac{R_s}{R_w}\right) + \left(\frac{k_R}{k_s} - 1\right)\ln\frac{R_s}{R_w}\right] = \frac{2\pi H}{\mu Q}(P_o - P_w)$$

$$\frac{2\pi k_R H(P_o - P_w)}{\mu Q} = \ln\frac{R_o}{R_w} + \left(\frac{k_R}{k_s} - 1\right)\ln\frac{R_s}{R_w}. \qquad (5.1.4)$$

Equation (5.1.4) is identical to Eq. (1.4.3), except for the second term on the right-hand side. This excess dimensionless pressure drop, due to the presence of the skin effect, is denoted by s, in which case Eq. (5.1.4) can be written as

$$\frac{2\pi k_R H(P_o - P_w)}{\mu Q} = \ln\frac{R_o}{R_w} + s. \qquad (5.1.5)$$

The effect of the skin is therefore to *add an additional component to the pressure drawdown*, over-and-above the drawdown that is due to the hydraulic resistance of the reservoir itself.

The *dimensional* form of the pressure drawdown in the well can be written as

$$P_w = P_o - \frac{\mu Q}{2\pi k_R H}\left[\ln\frac{R_o}{R_w} + s\right], \qquad (5.1.6)$$

$$\text{where} \quad s = \left(\frac{k_R}{k_s} - 1\right)\ln\frac{R_s}{R_w}. \qquad (5.1.7)$$

The additional pressure drop caused by the skin is

$$\Delta P_s = \frac{\mu Q s}{2\pi k_R H}. \tag{5.1.8}$$

Equation (5.1.7) shows that if the permeability in the damaged zone is reduced, or the thickness of the damaged zone is increased, the skin effect will increase. The value of the mechanical skin factor is usually less than 20. However, factors such as incomplete perforation of the well casing can lead to an "apparent skin factor" that may be as large as 300. On the other hand, well *stimulation* processes such as acidising can increase the near-wellbore permeability, and give rise to a *negative* skin factor. A practical lower limit to s is about -6.

The wellbore skin effect is illustrated schematically in Figure 5.2, adapted from *Pressure Transient Analysis* by Stanislav and Kabir (1990).

Another interpretation of the skin effect is based on the observation that the effect of skin is the same as the effect of having a smaller wellbore radius, i.e. it reduces the flow, for a given pressure drawdown. Hence, we can write Eq. (5.1.6) as

$$\begin{aligned} P_w &= P_o - \frac{\mu Q}{2\pi k_R H}\left[\ln\frac{R_o}{R_w} - \ln(e^{-s})\right] \\ &= P_o - \frac{\mu Q}{2\pi k_R H}\ln\left(\frac{R_o}{R_w e^{-s}}\right) \end{aligned}$$

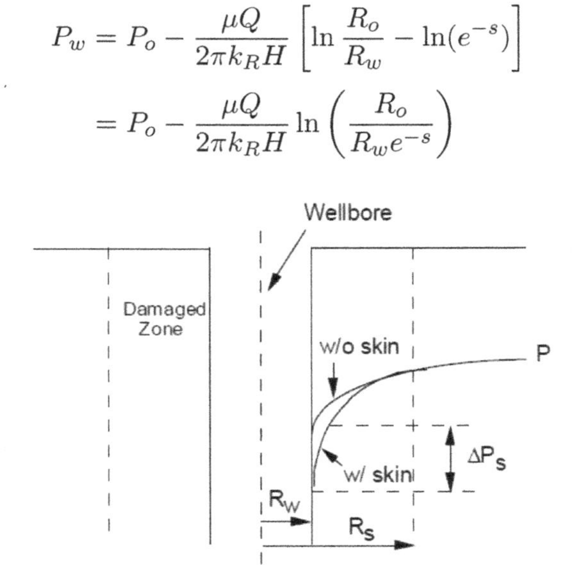

Figure 5.2. Effect of skin on the pressure drawdown near a well.

$$= P_o - \frac{\mu Q}{2\pi k_R H} \ln\left(\frac{R_o}{R_w^{\text{eff}}}\right), \tag{5.1.9}$$

$$\text{where} \quad R_w^{\text{eff}} = R_w e^{-s}. \tag{5.1.10}$$

The term R_w^{eff} can be thought of as the hypothetical wellbore radius, in an undamaged formation, that would lead to the same drawdown as is actually observed in the well that is surrounded by the damaged formation. However, this interpretation is potentially misleading, because in general it is *not* correct to merely replace the actual wellbore radius with the effective radius given in Eq. (5.1.10). This interpretation only applies to drawdown tests in an infinite reservoir.

5.2. Effect of Wellbore Skin on Drawdown or Buildup Tests

The preceding derivation was carried out under the steady-state assumption. However, the thickness of the skin region is usually small, and so transient effects due to the skin region will die out rapidly. The time required for "transient effects" to die out within the skin region can be estimated using Eq. (1.6.12), with the distance taken to be R_s:

$$t_s = \frac{(\phi\mu c)_s R_s^2}{4k_s}. \tag{5.2.1}$$

If we assume reasonable values of the parameters, such as $\phi \approx 10^{-1}$, $\mu \approx 10^{-3}\,\text{Pa s}$, $c \approx 10^{-9}/\text{Pa}$, $R_s \approx 10^0\,\text{m}$, and $k_s \approx 10^{-15}\,\text{m}^2$, we find that $t_s \approx 25\,\text{s}$. So, after a short period of time, the skin region will be in a quasi-steady regime. If the flow rate into the wellbore is constant, then the pressure drop through the skin region will be constant. Therefore, it is usually assumed that, even during a transient well test, the effect of the skin region is to contribute a constant, additional pressure drop in the wellbore, the magnitude of which is given by Eq. (5.1.8).

For example, the drawdown in an infinite reservoir producing at a constant rate, with a skin region around the wellbore, is found by adding the skin-related pressure drop given by Eq. (5.1.8), to the drawdown given by the line source solution:

$$P_w = P_i - \frac{\mu Q}{4\pi kH}\left[-Ei\left(\frac{\phi\mu c R_w^2}{4kt}\right) + 2s\right], \tag{5.2.2}$$

where we now revert to using k instead of k_R to denote the reservoir permeability, since the skin-zone permeability k_s has been absorbed into the skin factor, s.

If t is large enough that the logarithmic approximation can be used for the Ei function, the drawdown will be given by

$$P_w = P_i - \frac{\mu Q}{4\pi kH}\left[\ln\left(\frac{kt}{\phi\mu c R_w^2}\right) + 0.80907 + 2s\right], \tag{5.2.3}$$

$$\text{or,} \quad P_w = P_i - \frac{\mu Q}{4\pi kH}\ln\left(\frac{2.246e^{2s}kt}{\phi\mu c R_w^2}\right). \tag{5.2.4}$$

It is clear from Eq. (5.2.3) that the skin will have no effect on the semi-log slope of pressure versus $\ln t$, but it will shift the entire drawdown curve downward by a constant amount.

The skin factor can be found by proper interpretation of a pressure buildup test. Incorporating the skin effect into the equation for the buildup pressure, Eq. (3.2.5), gives

$$P_w = P_i - \frac{\mu Q}{4\pi kH}\left[\ln\left(\frac{2.246e^{2s}k(t+\Delta t)}{\phi\mu c R_w^2}\right) - \ln\left(\frac{2.246e^{2s}k\Delta t}{\phi\mu c R_w^2}\right)\right]$$

$$\rightarrow P_w = P_i - \frac{\mu Q}{4\pi kH}\ln\left(\frac{t+\Delta t}{\Delta t}\right). \tag{5.2.5}$$

The two skin terms cancel out, and do not appear in the equation for the wellbore pressure.

Now consider the difference between the wellbore pressure immediately before shut-in, which we will call P_w^-, and the wellbore pressure a short time after shut-in, P_w^+. Using Eq. (5.2.4) for P_w^-, and Eq. (5.2.5) for P_w^+, we find

$$\begin{aligned} P_w^+ - P_w^- &= -\frac{\mu Q}{4\pi kH}\ln\left(\frac{t+\Delta t}{\Delta t}\right) + \frac{\mu Q}{4\pi kH}\ln\left(\frac{2.246 e^{2s} kt}{\phi\mu c R_w^2}\right) \\ &= \frac{-\mu Q}{4\pi kH}\left[\ln\left(\frac{t+\Delta t}{t\Delta t}\right) - \ln\left(\frac{2.246k}{\phi\mu c R_w^2}\right) - 2s\right]. \quad (5.2.6) \end{aligned}$$

For times shortly after shut-in, Δt is small, and $(t + \Delta t)/t \approx 1$, so Eq. (5.2.6) reduces to

$$P_w^+ - P_w^- = \frac{\mu Q}{4\pi kH}\left[\ln \Delta t + \ln\left(\frac{2.246k}{\phi\mu c R_w^2}\right) + 2s\right], \quad (5.2.7)$$

which provides an equation that can be solved for s.

In practice, however, we cannot use the pressure at a time *immediately* after shut-in in Eq. (5.2.7), for several reasons:

(a) Wellbore storage effects, (called "afterproduction" in the case of a buildup test), will cause the actual pressure drawdown measured in the borehole at reservoir depth to deviate from that predicted by the equations used above, for a certain period of time; see Section 5.3 below.
(b) Equation (5.2.5) made use of the logarithmic approximation to the exponential integral function; this approximation is only valid after a sufficiently long time has elapsed after shut-in.

It is traditional practice to use Eq. (5.2.7) with P_w^+ evaluated after *one-hour* of shut-in time; this value was originally chosen because, when using "oilfield" units, $\ln \Delta t = \ln 1 = 0$ when $\Delta t = 1$ h. However, in many cases 1 h is *not* enough time for afterproduction to cease and radial flow to be established. Therefore, we must extrapolate the straight-line portion of the Horner plot back to $\Delta t = 1$ h, and use that pressure for P_w^+. Use of buildup data to estimate the skin factor is illustrated in more detail in the MSc course module on Well Test Analysis.

5.3. Wellbore Storage Phenomena

The pressure drawdown is usually measured at reservoir depth in the borehole, whereas the flow rate is usually measured at the wellhead. The flow rate Q in our equations, however, refers to the "sandface" flow rate from the reservoir into the borehole, denoted by $Q_{\rm sf}$. In a quasi-steady situation, the measured wellhead flow rate will be identical to the influx from the reservoir.

However, immediately after a change in flow rate, such as after the start of a drawdown test, or after flow is shut-in at the wellhead, these two flow rates will differ. The difference is due to the fact that the fluid in the borehole is compressing (or expanding) due to the changing pressure to which it is subjected.

For example, imagine that we begin to withdraw fluid at time $t = 0$ at rate $Q_{\rm wh}$ that is measured at the wellhead. Initially, the fluid that flows at the wellhead is taken mainly from the wellbore itself; only gradually does this fluid begin to be supplied by the reservoir. Eventually, the fluid in the wellbore reaches a quasi-steady state, and all of the flow measured at the wellhead does in fact emanate from the reservoir. After this time, $Q_{\rm wh} = Q_{\rm sf}$.

Conversely, if a well that is producing at a constant rate is shut-in at the surface, fluid will continue to enter the borehole from the reservoir, even though it is not allowed to flow out at the wellhead. This additional fluid will be trapped within the wellbore; as the mass of the fluid in the wellbore increases, the wellbottom pressure will gradually increase. These two situations are illustrated in Figure 5.3, modified from p. 43 of Stanislav and Kabir (1990).

We can begin to analyse the effect of wellbore storage by doing a mass balance on the fluid in the wellbore, along the lines of that given for the reservoir in Section 1.5. Assuming that the wellbore is completely filled with a single-phase liquid, then:

Mass flux in − Mass flux out = rate of change of mass storage

$$\text{i.e.} \qquad \rho_f Q_{\rm sf} - \rho_f Q_{\rm wh} = \frac{d(\rho_f V_w)}{dt}, \tag{5.3.1}$$

where V_w is the volume of the wellbore.

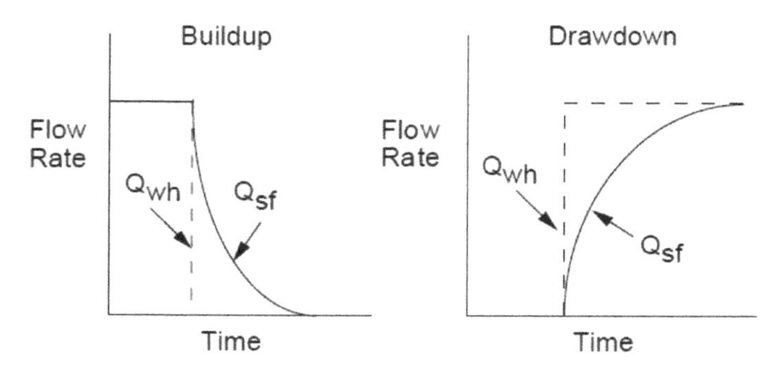

Figure 5.3. Effect of wellbore storage on pressure tests.

If we assume uniform pressure and uniform density of the fluid within the wellbore, we have

$$\rho_f(Q_{\text{sf}} - Q_{\text{wh}}) = V_w \frac{d\rho_f}{dt} = V_w \frac{d\rho_f}{dP_w}\frac{dP_w}{dt} = V_w \rho_f c_f \frac{dP_w}{dt},$$

$$\rightarrow Q_{\text{sf}} - Q_{\text{wh}} = V_w c_f \frac{dP_w}{dt} \equiv C_s \frac{dP_w}{dt}, \tag{5.3.2}$$

where $C_s = V_w c_f$ is the *wellbore storage coefficient*, and V_w is the total volume of fluid within the wellbore, from the wellbottom to the surface.

Aside: If the wellbore is not fully filled with liquid, the appropriate expression for C_s can be found in the SPE monograph *Advances in Well Test Analysis* (Earlougher, 1977); see also Problem 5.1.

At early times after the start of a drawdown test, $Q_{\text{wh}} = Q$, where Q is the nominal flow rate, but $Q_{\text{sf}} \cong 0$, so Eq. (5.3.2) takes the form

$$-Q = C_s \frac{dP_w}{dt}. \tag{5.3.3}$$

Equation (5.3.3) can be integrated, using the initial condition that $P_w = P_i$ when $t = 0$:

$$-\int_0^t \frac{Q}{C_s} dt = \int_{P_i}^{P_w(t)} dP_w \quad \rightarrow \quad P_w = P_i - \frac{Qt}{C_s}. \tag{5.3.4}$$

Equation (5.3.4) can be written in dimensionless form as follows. First, rewrite Eq. (5.3.4) in terms of the dimensionless pressure:

$$P_{\rm Dw} = \frac{2\pi kH(P_i - P_w)}{\mu Q} = \frac{2\pi kHQt}{\mu QC_s}. \tag{5.3.5}$$

Now, express the right-hand side of Eq. (5.3.5) in terms of $t_{\rm Dw}$:

$$P_{\rm Dw} = \frac{2\pi kHQ}{\mu QC_s}\frac{\phi\mu c_t R_w^2 t_{\rm Dw}}{k} \equiv \frac{t_{\rm Dw}}{C_D}, \tag{5.3.6}$$

where the *dimensionless wellbore storage coefficient* C_D is defined by

$$C_D = \frac{C_s}{2\pi H\phi c_t R_w^2}, \tag{5.3.7}$$

where c_t is the total reservoir compressibility. Aside from the factor of 2, C_D is the ratio of the storativity of the fluid in the *entire* borehole, to the storativity of the rock that had occupied the borehole (in the reservoir, *not* including the overburden) before the borehole was drilled.

On a log-log plot, Eq. (5.3.6) takes the form

$$\ln P_{\rm Dw} = \ln t_{\rm Dw} - \ln C_D, \quad \rightarrow \frac{d\ln P_{\rm Dw}}{d\ln t_{\rm Dw}} = 1. \tag{5.3.8}$$

Therefore, a slope of unity occurring at early times on a log-log plot (dimensionless or not) of wellbore pressure versus time is an indication of a pressure response that is *dominated by wellbore-storage effects*, and *not* by the reservoir properties!

5.4. Effect of Wellbore Storage on Well Tests

The preceding analysis focused on fluid produced from wellbore storage, and neglected flow from the reservoir into the wellbore. In reality, fluid is supplied to the wellhead by wellbore storage at early times, but these effects gradually die out as the flow regime in the wellbore reaches a steady state.

An analysis that includes contributions to the measured well-head flow rate from both wellbore storage *and* the reservoir, can be carried out by solving Eq. (2.1.1), with the inner boundary condition, Eq. (2.1.3), replaced by (see Eq. (5.3.2)):

$$\left(\frac{2\pi kH}{\mu}R\frac{\partial P}{\partial R}\right)_{R_w} = Q_{\text{wh}} + C_s\left(\frac{\partial P}{\partial t}\right)_{R_w}. \tag{5.4.1}$$

Note that the right-hand side is the sand-face flow rate, Q_{sf}, which is the flow rate that must be used in Darcy's law for the reservoir.

This problem has been solved by Wattenbarger and Ramey (1970) using finite differences, and by Agarwal *et al.* (1970) using Laplace transforms. Their analyses show that there are three regimes of behaviour for a well in an infinite reservoir, with skin and wellbore storage effects:

(a) An early-time regime that is dominated by wellbore storage, during which the drawdown is given by Eq. (5.3.6). The duration of this regime is defined by

$$t_{\text{Dw}} < C_D(0.04 + 0.02s). \tag{5.4.2}$$

(b) A transition regime during which both wellbore storage and reservoir inflow contribute to the wellhead flow rate. The duration of this regime is defined by

$$C_D(0.04 + 0.02s) < t_{\text{Dw}} < C_D(60 + 3.5s). \tag{5.4.3}$$

(c) A third regime during which wellbore storage effects have died off, and the drawdown is given by Eq. (5.2.4). The duration of this regime is defined by

$$t_{\text{Dw}} > C_D(60 + 3.5s). \tag{5.4.4}$$

Criterion (5.4.4) can be used to estimate when the drawdown curve can be approximated by the logarithmic equation (5.2.4). During a buildup test, however, the time required in order to reach a "straight line" on the Horner plot is best estimated by the following equation (Chen and Brigham, 1978)

$$t_{\text{Dw}} > 50C_D\exp(-0.14s). \tag{5.4.5}$$

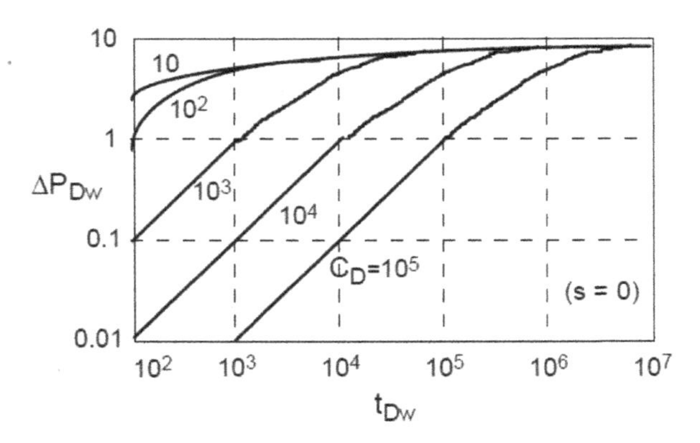

Figure 5.4. Effect of wellbore storage on drawdown.

The effect of wellbore storage on the pressure drawdown in the well is shown in very rough schematic form in Figure 5.4.

Finally, it should be noted that wellbore storage effects during buildup tests can be eliminated by having the flow shut-in at the reservoir depth, rather than at the wellhead.

Problem for Chapter 5

Problem 5.1. Imagine a wellbore that is filled with liquid only up to some height h above the top of the reservoir. The liquid has density ρ. Even if this liquid were incompressible, a wellbore storage effect would still occur, due to the raising or lowering of the fluid column. Derive an expression for the wellbore storage coefficient C_s in this case.

Chapter 6

Production From Bounded Reservoirs

In this chapter, we will solve the pressure diffusion equation in finite reservoirs that are bounded by impermeable no-flow boundaries or by aquifers that provide constant-pressure boundaries. We start with a well at the centre of a circular reservoir, and then move on to consider reservoirs having more general shapes. The Boltzmann transformation that was used to develop the line source solution does not work in a bounded reservoir, and so we will use the method of eigenfunction expansions to relate the pressures and flow rates observed at the wellbore to physical parameters such as permeability, storativity and reservoir size/shape.

6.1. Production From Bounded Reservoirs and/or Finite Drainage Areas

Thus far, we have assumed that our wells are located in an infinite, or at least semi-infinite, reservoir. In reality, the reservoir is always of finite size, and the pressure disturbance will eventually reach some outer boundary that effectively encloses the reservoir. This outer boundary may be an "aquifer", i.e. a vast expanse of water-filled rock that surrounds the hydrocarbon-filled reservoir. In this case, the appropriate outer boundary condition may be one of constant pressure.

On the other hand, if a number of wells are producing from the same reservoir, each will drain fluid from only a finite region, and so each well will effectively behave as if it were surrounded by a no-flow

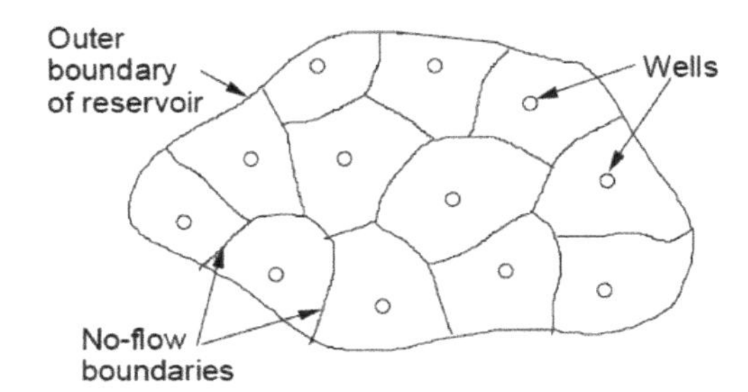

Figure 6.1. Bounded reservoir containing several production wells.

boundary, as illustrated in Figure 6.1. The effective drainage area of each well will depend on the production rate of that well, and of all nearby wells (Dake, 1978).

Problems involving a well producing oil from a finite, bounded drainage area can be solved using either of two mathematical techniques: Laplace transforms, or eigenfunction expansions. Laplace transforms will be discussed in Chapter 7. In this chapter, we will use the method of eigenfunction expansions to solve the problem of a well located at the centre of a circular reservoir, with its outer boundary maintained at constant pressure, and with a constant wellbore pressure. This is not the most important example of a bounded-reservoir problem, but it will allow us to demonstrate the method of eigenfunction expansions in a (relatively) simple context.

The first point to note is that the Boltzmann transformation, $\eta = \phi\mu cR^2/kt$, *is not applicable in bounded-reservoir problems.* The reason that the Boltzmann transformation "worked" in the infinite-reservoir case is that the initial condition, at $t = 0$, and the outer boundary condition, at $R \to \infty$, both correspond to $\eta \to \infty$, and both also correspond to the initial pressure, P_i. This "collapse" of two conditions into one allowed us to replace a second-order partial differential equation (PDE), which requires two boundary conditions + one initial condition = three conditions, with a second-order ordinary differential equation (ODE), which requires a total of only two boundary conditions.

The collapse of the initial condition and the far-field boundary condition into one single boundary condition does not occur in the bounded-reservoir problem, because when $R = R_e$, η still varies with time, and so does not correspond to any constant value at which a boundary condition can be imposed. More generally, a Boltzmann-type transformation will *not* work for a problem that contains a natural physical length scale, such as a wellbore radius or an outer radius; the problem of an infinitely small line source in an infinite reservoir had no natural length scale associated with it.

6.2. Well at the Centre of a Circular Reservoir With Constant Pressure on Its Outer Boundary and Constant Wellbore Pressure

Imagine that we have a well located at the centre of a circular reservoir, which is initially at some initial pressure P_i. At time $t = 0$, the pressure in the wellbore is immediately lowered to some value P_w, and it is thereafter maintained at that value (Figure 6.2). In contrast to previous problems, in which the flow rate was controlled, and we calculated the drawdown, in this problem the drawdown is specified, and we must calculate the flow rate. This problem can be stated mathematically as follows:

$$\text{Governing PDE:} \quad \frac{1}{R}\frac{d}{dR}\left(R\frac{dP}{dR}\right) = \frac{\phi\mu c}{k}\frac{dP}{dt}, \tag{6.2.1}$$

$$\text{BC at wellbore:} \quad P(R = R_w, t) = P_w, \tag{6.2.2}$$

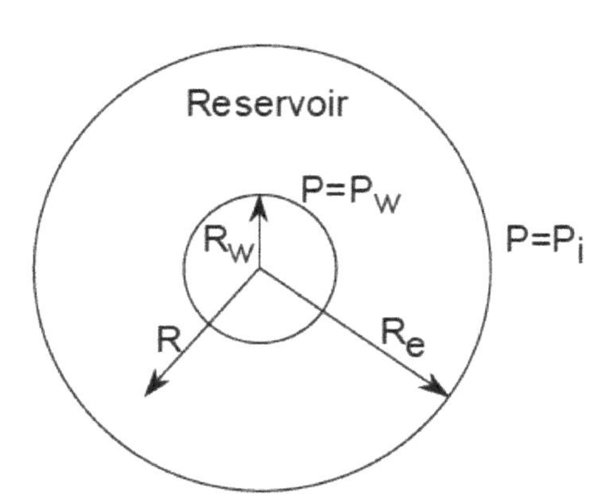

Figure 6.2. Circular reservoir with constant-pressure inner and outer boundaries.

$$\text{BC at outer boundary:} \quad P(R = R_e, t) = P_i, \tag{6.2.3}$$

$$\text{Initial condition:} \quad P(R, t = 0) = P_i. \tag{6.2.4}$$

The first step in solving this problem is to simplify its appearance by introducing dimensionless variables. Define

$$\text{Dimensionless radius:} \quad R_D = R/R_w, \tag{6.2.5}$$

$$\text{Dimensionless time:} \quad t_D = \frac{kt}{\phi \mu c R_w^2}, \tag{6.2.6}$$

$$\text{Dimensionless pressure:} \quad P_D = \frac{P_i - P}{P_i - P_w}. \tag{6.2.7}$$

The dimensionless time is defined here similarly to that used in the line source solution, with the well radius as the length scale. (To simplify the notation, in this chapter we will usually write t_D. But in some cases, to show the connection with the infinite-reservoir problem, we will write the dimensionless time as t_{Dw}.) The dimensionless pressure is defined so that it equals 1 at the well, and always equals 0 at the outer boundary. Note that the production rate Q is *a priori* unknown, so we cannot define P_D as we did before, in terms of Q. In terms of these dimensionless variables, Eqs. (6.2.1–6.2.4) take the form

$$\text{Governing PDE:} \quad \frac{1}{R_D}\frac{d}{dR_D}\left(R_D\frac{dP_D}{dR_D}\right) = \frac{dP_D}{dt_D}, \tag{6.2.8}$$

$$\text{BC at wellbore:} \quad P_D(R_D = R_w/R_w = 1, t_D) = 1, \tag{6.2.9}$$

$$\text{BC at outer boundary:} \quad P_D(R_D = R_e/R_w \equiv R_{\mathrm{De}}, t_D) = 0, \tag{6.2.10}$$

$$\text{Initial condition:} \quad P_D(R_D, t_D = 0) = 0. \tag{6.2.11}$$

Next, we invoke the principle of superposition to break up the pressure into a steady-state part, $P_D^s(R_D)$, and a transient part, $p_D(R_D, t_D)$:

$$P_D(R_D, t_D) = P_D^s(R_D) + p_D(R_D, t_D). \tag{6.2.12}$$

By definition, the steady-state pressure function must satisfy Eq. (6.2.8), as well as both BCs, (6.2.9) and (6.2.10). The time derivative is zero for the steady-state part, by definition, so the governing differential equation for $P_D^s(R_D)$ is Eq. (6.2.8), with the right-hand side set to zero. Therefore, as time is no longer relevant for the steady-state pressure, $P_D^s(R_D)$ is governed by the following ODE:

$$\text{Governing ODE:} \qquad \frac{1}{R_D}\frac{d}{dR_D}\left(R_D\frac{dP_D^s}{dR_D}\right) = 0, \tag{6.2.13}$$

$$\text{BC at wellbore:} \qquad P_D^s(R_D = 1) = 1, \tag{6.2.14}$$

$$\text{BC at outer boundary:} \qquad P_D^s(R_D = R_{\text{De}}) = 0. \tag{6.2.15}$$

To find the steady-state pressure, we first integrate Eq. (6.2.13) once, using an *indefinite* integral, to obtain

$$R_D\frac{dP_D^s}{dR_D} = \text{constant} = B. \tag{6.2.16}$$

Next, we separate the variables in Eq. (6.2.16) and integrate again:

$$\int^{P_D^s(R_D)} dP_D^s = \int^{R_D} B\frac{dR_D}{R_D} + A$$

$$\to P_D^s(R_D) = B\ln R_D + A, \tag{6.2.17}$$

where A and B are constants of integration. Comparison of Eq. (6.2.17) with the boundary conditions (6.2.14) and (6.2.15) shows that $A = 1$ and $B = -1/\ln R_{\text{De}}$, and so Eq. (6.2.17) can be written as

$$P_D^s(R_D) = 1 - \frac{\ln R_D}{\ln R_{\text{De}}} = \frac{-\ln(R_D/R_{\text{De}})}{\ln R_{\text{De}}}. \tag{6.2.18}$$

Note 1: The steady-state pressure satisfies the diffusion equation (6.2.8), and the boundary conditions (6.2.9, 6.2.10), but it *does not* satisfy the initial condition (6.2.11); this is why we also need the transient component of the pressure.

Note 2: The steady-state pressure given by Eq. (6.2.18) is just the Thiem equation, Eq. (1.4.3), in dimensionless form!

If we now substitute Eq. (6.2.12) into Eqs. (6.2.8–6.2.11), and make use of Eq. (6.2.18), we see that the transient pressure function must satisfy the following equations:

Governing PDE: $$\frac{1}{R_D}\frac{d}{dR_D}\left(R_D\frac{dp_D}{dR_D}\right)=\frac{dp_D}{dt_D}, \tag{6.2.19}$$

BC at wellbore: $$p_D(R_D=1,t_D)=0, \tag{6.2.20}$$

BC at outer boundary: $$p_D(R_D=R_{\mathrm{De}},t_D)=0, \tag{6.2.21}$$

Initial condition: $$p_D(R_D,t_D=0)=\frac{\ln(R_D/R_{\mathrm{De}})}{\ln R_{\mathrm{De}}}. \tag{6.2.22}$$

Thus far, we have transformed a diffusion equation with a non-zero BC, Eq. (6.2.9), and zero IC, Eq. (6.2.11), into a diffusion equation with zero BC and a non-zero IC, Eq. (6.2.22). We may not appear to be making progress, but in fact we are because the eigenfunction method requires "zero" boundary conditions, but does not require "zero" initial conditions.

To solve Eqs. (6.2.19–6.2.22), we again make use of the superposition principle, and first search for as many functions as we can find that *each* satisfy Eq. (6.2.19) and the BCs (6.2.20, 6.2.21); later, we will superpose these functions to satisfy the initial condition, Eq. (6.2.22). We proceed as follows:

(a) Assume that these functions can be written in the form

$$p_D(R_D,t_D)=F(R_D)G(t_D). \tag{6.2.23}$$

(b) Insert Eq. (6.2.23) into Eq. (6.2.19) to find

$$G(t_D)\left[\frac{1}{R_D}\frac{d}{dR_D}\left(R_D\frac{dF}{dR_D}\right)\right]=F(R_D)\frac{dG}{dt_D}. \tag{6.2.24}$$

(c) Divide through by $F(R_D)G(t_D)$ to find

$$\frac{1}{R_DF(R_D)}\frac{d}{dR_D}\left(R_D\frac{dF}{dR_D}\right)=\frac{1}{G(t_D)}\frac{dG}{dt_D}. \tag{6.2.25}$$

(d) The left-hand side of Eq. (6.2.25) does not depend on t_D, and the right-hand side does not depend on R_D; since they are equal to each other, they cannot depend on t_D or on R_D! Hence, each side of Eq. (6.2.25) must equal a *constant*, which we call $-\lambda^2$:

$$\frac{1}{R_D F(R_D)} \frac{d}{dR_D}\left(R_D \frac{dF}{dR_D}\right) = \frac{1}{G(t_D)} \frac{dG}{dt_D} = -\lambda^2. \qquad (6.2.26)$$

(e) The space-dependent part of Eq. (6.2.26) can be written as

$$F''(R_D) + \frac{1}{R_D} F'(R_D) + \lambda^2 F(R_D) = 0. \qquad (6.2.27)$$

(f) We now make a change of variables $x = \lambda R_D$, after which Eq. (6.2.27) can be written as

$$F''(x) + \frac{1}{x} F'(x) + F(x) = 0. \qquad (6.2.28)$$

Equation (6.2.28) is known as a *Bessel equation of order zero.* It is a second-order ordinary differential equation, so it must have two independent solutions. One solution is easily found by assuming a power-series solution:

$$F(x) = \sum_{n=0}^{\infty} a_n x^n. \qquad (6.2.29)$$

If we insert Eq. (6.2.29) into Eq. (6.2.28), we find

$$\sum_{n=0}^{\infty} n(n-1) a_n x^{n-2} + \sum_{n=0}^{\infty} n a_n x^{n-2} + \sum_{n=0}^{\infty} a_n x^n = 0. \qquad (6.2.30)$$

In order to have x appear to the nth power in each term, so that we can easily add them together, we let $n \to n + 2$ in the first two series (which is permissible, because n is just a "dummy index"), which, after combining the first two series and taking the "unmatched" terms outside of the series, leads to

$$0a_0 x^{-2} + a_1 x^{-1} + \sum_{n=0}^{\infty} [(n+2)^2 a_{n+2} + a_n] x^n = 0. \qquad (6.2.31)$$

The coefficient of *each* power of x in Eq. (6.2.31) must be zero. The x^{-2} term already has a pre-factor of zero, so it will vanish regardless of the choice of a_0; we therefore may as well pick

$$a_0 = 1. \tag{6.2.32}$$

In order for the x^{-1} term to vanish, we *must* pick

$$a_1 = 0. \tag{6.2.33}$$

For all higher-order terms to vanish, Eq. (6.2.31) shows that the coefficients must satisfy the following *recursion relationship*

$$a_{n+2} = \frac{-a_n}{(n+2)^2}. \tag{6.2.34}$$

This recursion relationship allows us to generate all subsequent coefficients, starting with the first two coefficients, which are given by Eqs. (6.2.32) and (6.2.33). For example, with $n = 0$, Eq. (6.2.34) generates $a_2 = -a_0/4 = -1/4$. In this manner, we find:

$$a_3 = a_5 = a_7 = \cdots = 0, \tag{6.2.35}$$

$$a_2 = \frac{-1}{4}, \quad a_4 = \frac{1}{64}, \quad a_6 = \frac{-1}{2304} \cdots. \tag{6.2.36}$$

The solution to Eq. (6.2.28) that is given by Eqs. (6.2.29), (6.2.35) and (6.2.36) is known as the *Bessel function of the first kind, of order zero*, and is denoted by $J_0(x)$:

$$J_0(x) = 1 - \frac{x^2}{4} + \frac{x^4}{64} - \frac{x^6}{2304} + \cdots,$$

$$\text{or} \quad J_0(x) = 1 - \frac{x^2}{2^2(1!)^2} + \frac{x^4}{2^4(2!)^2} - \frac{x^6}{2^6(3!)^2} + \cdots, \tag{6.2.37}$$

where $n! \equiv 1 \times 2 \times 3 \times \cdots \times n$ is the *factorial* function.

The function $J_0(x)$ is similar to a damped cosine function (see Figure 6.3.). It "starts" at $J_0(0) = 1$, oscillates, but not quite periodically, and then slowly decays to zero according to a factor proportional to $1/\sqrt{x}$. Specifically (see *A Treatise on the*

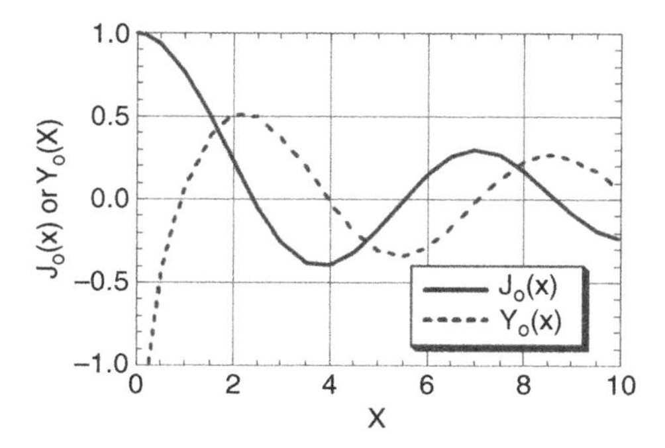

Figure 6.3. Bessel functions of order zero.

Theory of Bessel Functions, Watson (1944), or almost any book on advanced applied mathematics), for large values of x, this function is approximately given by

$$J_0(x) \approx \sqrt{\frac{2}{\pi x}} \cos\left(x - \frac{\pi}{4}\right) \quad as \quad x \to \infty. \tag{6.2.38}$$

The derivation of the second independent solution to Eq. (6.2.28) requires a somewhat lengthier procedure (which will not be given here) because this second solution is not analytic, i.e. it is not a pure power series. This solution, called the *Bessel function of the second kind, of order zero*, is defined by

$$Y_0(x) = \frac{2\ln(\gamma x/2)}{\pi} J_0(x) - \frac{2}{\pi} \sum_{n=1}^{\infty} \frac{(-1)^n h_n}{(n!)^2} \left(\frac{x}{2}\right)^{2n}, \tag{6.2.39}$$

where $\gamma \approx 1.781$, and

$$h_n = 1 + \frac{1}{2} + \cdots + \frac{1}{n}. \tag{6.2.40}$$

This function becomes infinitely negative as $x \to 0$, due to the $\ln x$ term, then oscillates in a manner similar to $J_0(x)$, and eventually decays to zero according to (Figure 6.3)

$$Y_0(x) \approx \sqrt{\frac{2}{\pi x}} \sin\left(x - \frac{\pi}{4}\right) \quad \text{as} \quad x \to \infty. \tag{6.2.41}$$

The general solution to Eq. (6.2.28) can be written as a linear combination of these two kinds of Bessel functions:

$$F(x) = AJ_0(x) + BY_0(x), \tag{6.2.42}$$

where these two constants A and B are unrelated to the constants that appeared in Eq. (6.2.17). Recalling that $x = \lambda R_D$, we can say that the function

$$F(R_D) = AJ_0(\lambda R_D) + BY_0(\lambda R_D) \tag{6.2.43}$$

will be a solution to Eq. (6.2.28) for any value of λ. (Recall that the general solution to a second-order ODE contains two arbitrary constants, in this case A and B.)

However, Eq. (6.2.43) will satisfy the boundary conditions (6.2.20) and (6.2.21) only for *certain special values* of λ, which we will now find. Insertion of Eq. (6.2.43) into BCs (6.2.20) and (6.2.21) yields

$$AJ_0(\lambda) + BY_0(\lambda) = 0, \tag{6.2.44}$$

$$AJ_0(\lambda R_{\mathrm{De}}) + BY_0(\lambda R_{\mathrm{De}}) = 0. \tag{6.2.45}$$

These two equations can be written in matrix form as

$$\begin{bmatrix} J_0(\lambda) & Y_0(\lambda) \\ J_0(\lambda R_{\mathrm{De}}) & Y_0(\lambda R_{\mathrm{De}}) \end{bmatrix} \begin{bmatrix} A \\ B \end{bmatrix} = \begin{bmatrix} 0 \\ 0 \end{bmatrix}. \tag{6.2.46}$$

In order for Eq. (6.2.46) to have non-zero solutions for A and B, the determinant of the matrix must be *zero*, i.e.

$$J_0(\lambda)Y_0(\lambda R_{\mathrm{De}}) - Y_0(\lambda)J_0(\lambda R_{\mathrm{De}}) = 0. \tag{6.2.47}$$

The values of λ that satisfy Eq. (6.2.47) are called the *eigenvalues* of this problem. They will depend on the dimensionless size of the reservoir, $R_{\mathrm{De}} = R_e/R_w$. It can be proven that there are an infinite number of eigenvalues, and they can be arranged in order as

$$0 < \lambda_1 < \lambda_2 < \cdots < \lambda_n \to \infty. \tag{6.2.48}$$

Each of these eigenvalues generates its own eigenfunction, Eq. (6.2.43):

$$F_n(R_D) = A_n J_0(\lambda_n R_D) + B_n Y_0(\lambda_n R_D). \tag{6.2.49}$$

From Eq. (6.2.44), we see that A_n and B_n are related by

$$B_n = \frac{-A_n J_0(\lambda_n)}{Y_0(\lambda_n)}, \tag{6.2.50}$$

and so Eq. (6.2.49) can be written as

$$\begin{aligned} F_n(R_D) &= A_n J_0(\lambda_n R_D) - A_n \frac{J_0(\lambda_n)}{Y_0(\lambda_n)} Y_0(\lambda_n R_D) \\ &= \frac{A_n}{Y_0(\lambda_n)}[Y_0(\lambda_n)J_0(\lambda_n R_D) - J_0(\lambda_n)Y_0(\lambda_n R_D)] \\ &= C_n[Y_0(\lambda_n)J_0(\lambda_n R_D) - J_0(\lambda_n)Y_0(\lambda_n R_D)], \end{aligned} \tag{6.2.51}$$

where C_n is just another name for $A_n/Y_0(\lambda_n)$, and the functions inside the brackets are called the *eigenfunctions* of this problem. Note that at this stage, we do not yet know the values of the constants C_n.

We now return to the time-dependent part of the solution, which, according to Eq. (6.2.26), must satisfy

$$\frac{dG_n}{dt_D} = -\lambda_n^2 G_n(t_D). \tag{6.2.52}$$

The solution to Eq. (6.2.52) is

$$G_n(t_D) = e^{-\lambda_n^2 t_D}. \tag{6.2.53}$$

Note that we do not need an arbitrary constant of integration for $G_n(t_D)$, since the function $F_n(R_D)$ already includes an arbitrary constant, as seen in Eq. (6.2.51), and eventually these two functions will be multiplied together.

Recalling Eq. (6.2.23), the most general solution that satisfies the diffusion equation (6.2.19) and the BCs (6.2.20, 6.2.21) is given by

$$p_D(R_D, t_D) = \sum_{n=1}^{\infty} C_n[Y_0(\lambda_n)J_0(\lambda_n R_D) - J_0(\lambda_n)Y_0(\lambda_n R_D)]e^{-\lambda_n^2 t_D}. \tag{6.2.54}$$

All that remains now is to satisfy the initial condition, Eq. (6.2.22). Evaluating Eq. (6.2.54) at $t_D = 0$, and invoking Eq. (6.2.22), yields

$$\sum_{n=1}^{\infty} C_n[Y_0(\lambda_n)J_0(\lambda_n R_D) - J_0(\lambda_n)Y_0(\lambda_n R_D)] = \frac{\ln(R_D/R_{\text{De}})}{\ln R_{\text{De}}}. \tag{6.2.55}$$

The constants C_n must be chosen so that Eq. (6.2.55) is satisfied for all values of R_D. This can be done using the orthogonality properties of eigenfunctions, as is shown in Chapter 10 of *The Flow of Homogeneous Fluids through Porous Media* (Muskat, 1937). The result, skipping the details, is

$$C_n = \frac{\pi J_0(\lambda_n)J_0(\lambda_n R_{\text{De}})}{J_0^2(\lambda_n) - J_0^2(\lambda_n R_{\text{De}})}. \tag{6.2.56}$$

We now combine the transient and steady-state pressures to find, from Eqs. (6.2.12), (6.2.18), (6.2.54) and (6.2.56), the full expression for the pressure as a function of R and t:

$$P_D(R_D, t_D) = \frac{-\ln(R_D/R_{\text{De}})}{\ln R_{\text{De}}} + \sum_{n=1}^{\infty} \frac{\pi J_0(\lambda_n)J_0(\lambda_n R_{\text{De}})}{J_0^2(\lambda_n) - J_0^2(\lambda_n R_{\text{De}})} U_n(\lambda_n R_D)e^{-\lambda_n^2 t_D}, \tag{6.2.57}$$

where $U_n(\lambda_n R_D) = Y_0(\lambda_n)J_0(\lambda_n R_D) - J_0(\lambda_n)Y_0(\lambda_n R_D)$.

This completes the solution to this problem. The *dimensional* values can be found by recalling Eqs. (6.2.5–6.2.7).

The flow rate into the well can be found by differentiating the pressure, and applying Darcy's law at the wellbore. A detailed analysis of the flow rate would show that at early times it is proportional to $t^{-1/2}$, and then gradually approaches a steady-state value given by the Thiem equation, Eq. (1.4.3).

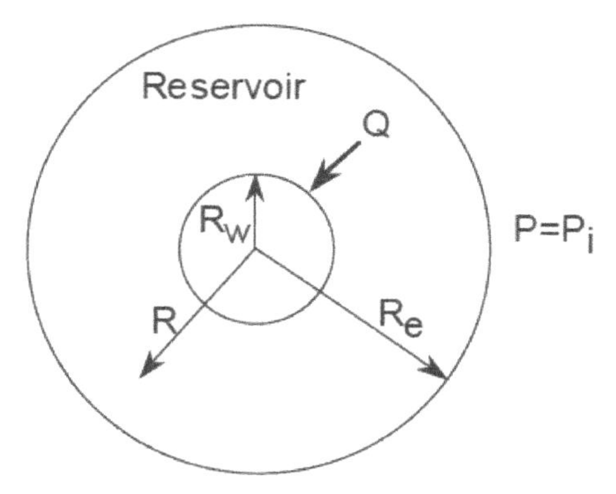

Figure 6.4. Circular reservoir with constant flow rate into well, constant pressure at outer boundary.

6.3. Well at the Centre of a Circular Reservoir With Constant Pressure on Its Outer Boundary and Constant Flow Rate into the Wellbore

Consider a well at the centre of a circular reservoir, producing fluid at a constant rate, with the pressure at the outer boundary maintained at the initial pressure at all times (Figure 6.4). The mathematical formulation of this problem is equivalent to that discussed in Section 6.2, except that the inner BC, Eq. (6.2.2) is replaced by

$$\text{BC at wellbore:} \quad \left(\frac{2\pi kH}{\mu} R\frac{dP}{dR}\right)_{R=R_w} = Q, \tag{6.3.1}$$

where we take $Q > 0$ if the fluid flows *into* the borehole.

This problem can also be solved using the method of eigenfunction expansions, as in the previous section. The solution is (Muskat, 1937, p. 643)

$$\Delta P_D(R_D, t_D) = -\ln\left(\frac{R_D}{R_{\mathrm{De}}}\right) - \sum_{n=1}^{\infty} \frac{\pi J_0^2(\lambda_n R_{\mathrm{De}}) U_n(\lambda_n R_D)}{\lambda_n [J_0^2(\lambda_n R_{\mathrm{De}}) - J_1^2(\lambda_n)]} e^{-\lambda_n^2 t_D}, \tag{6.3.2}$$

where the eigenfunctions U_n are given by

$$U_n(\lambda_n R_D) = Y_1(\lambda_n) J_0(\lambda_n R_D) - J_1(\lambda_n) Y_0(\lambda_n R_D), \tag{6.3.3}$$

and the eigenvalues λ_n are defined implicitly by

$$U_n(\lambda_n R_{\rm De}) = Y_1(\lambda_n)J_0(\lambda_n R_{\rm De}) - J_1(\lambda_n)Y_0(\lambda_n R_{\rm De}) = 0. \qquad (6.3.4)$$

The functions J_1 and Y_1 are *Bessel functions of order one*, of the first and second kind, respectively, and are defined by

$$J_1(x) \equiv -\frac{dJ_0(x)}{dx}, \quad Y_1(x) \equiv -\frac{dY_0(x)}{dx}. \qquad (6.3.5)$$

The dimensionless variables in Eq. (6.3.2) are defined by

$$\text{Dimensionless radius:} \quad R_D = R/R_w, \qquad (6.3.6)$$

$$\text{Dimensionless time:} \quad t_D = \frac{kt}{\phi\mu c R_w^2}, \qquad (6.3.7)$$

$$\text{Dimensionless drawdown:} \quad \Delta P_D = \frac{2\pi kH(P_i - P)}{\mu Q}. \qquad (6.3.8)$$

The pressure in the wellbore, $\Delta P_D(t_D)$, is found by setting $R_D = 1$ in Eq. (6.3.2), and then making use of the following Bessel-function identity (Muskat, 1937, p. 631):

$$Y_1(x)J_0(x) - J_1(x)Y_0(x) = \frac{-2}{\pi x}. \qquad (6.3.9)$$

The result is (Matthews and Russell, 1967, p. 12)

$$\Delta P_{\rm Dw}(t_D) = \ln R_{\rm De} + \sum_{n=1}^{\infty} \frac{2J_0^2(\lambda_n R_{\rm De})e^{-\lambda_n^2 t_D}}{\lambda_n^2[J_0^2(\lambda_n R_{\rm De}) - J_1^2(\lambda_n)]}. \qquad (6.3.10)$$

A detailed (but mathematically cumbersome) analysis of this solution would show that at early times, the pressure agrees with that given by the Theis line source solution. This is to be expected because at early times the pressure pulse will not yet have reached the outer boundary of the reservoir, and so the finite-reservoir solution should coincide with the infinite-reservoir solution. Eventually, the wellbore reaches a steady-state pressure (see Figure 6.6) that is equivalent to that which occurs in the Thiem problem.

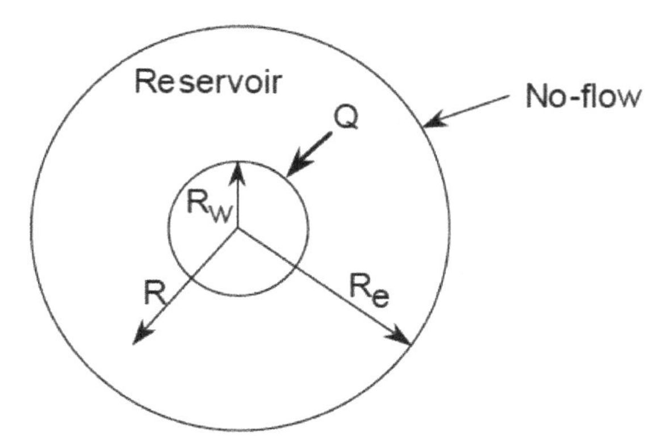

Figure 6.5. Circular reservoir with constant flow rate into borehole, and no flow across the outer boundary.

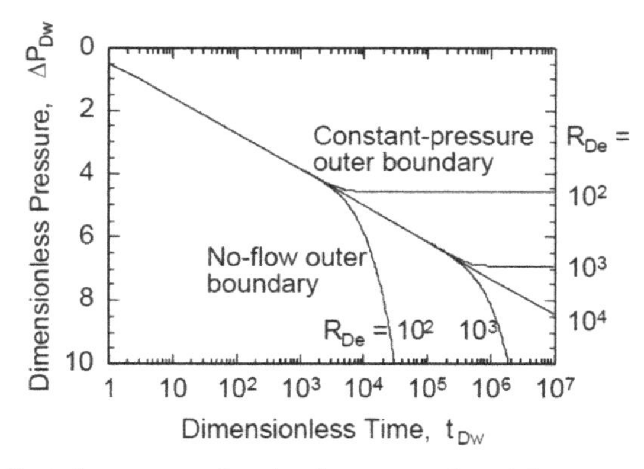

Figure 6.6. Well at the centre of a circular reservoir, with constant flow rate into the wellbore.

6.4. Well at the Centre of a Circular Reservoir With a No-flow Outer Boundary and Constant Flow Rate into the Wellbore

The problem of a well at the centre of a circular reservoir with a no-flow outer boundary and constant flow rate into the wellbore (Figure 6.5) is similar to the problem treated in Section 6.3, except that the *outer* BC is now replaced by

$$\text{BC at outer boundary:} \quad \left(\frac{2\pi kH}{\mu} R \frac{dP}{dR}\right)_{R=R_e} = 0. \qquad (6.4.1)$$

The solution to this problem was found by Muskat (1937) using the eigenfunction method, and was re-derived by van Everdingen and Hurst (1949), using Laplace transforms (see Chapter 7). The solution is

$$\Delta P_D(R_D, t_D) = \frac{1}{R_{\rm De}^2 - 1}\left[\frac{R_D^2}{2} + 2t_D - R_{\rm De}^2 \ln R_D\right] - \left[\frac{3R_{\rm De}^4 - 4R_{\rm De}^4 \ln R_{\rm De} - 2R_{\rm De}^2 - 1}{4(R_{\rm De}^2 - 1)^2}\right] + \sum_{n=1}^{\infty} \frac{\pi J_1^2(\lambda_n R_{\rm De}) U_n(\lambda_n R_D)}{\lambda_n [J_1^2(\lambda_n R_{\rm De}) - J_1^2(\lambda_n)]} e^{-\lambda_n^2 t_D}, \qquad (6.4.2)$$

where the eigenfunctions U_n are given by

$$U_n(\lambda_n R_D) = J_1(\lambda_n) Y_0(\lambda_n R_D) - Y_1(\lambda_n) J_0(\lambda_n R_D), \qquad (6.4.3)$$

the eigenvalues λ_n are defined implicitly by

$$J_1(\lambda_n) Y_1(\lambda_n R_{\rm De}) - Y_1(\lambda_n) J_1(\lambda_n R_{\rm De}) = 0, \qquad (6.4.4)$$

and the dimensionless variables are defined as in Section 6.3.

Because of the importance of this problem, we will examine this solution in detail. The pressure in the wellbore is found by setting $R_D = 1$ in Eq. (6.4.2), and again using Eq. (6.3.9) to simplify

$$\Delta P_{\rm Dw}(t_D) = \frac{1}{R_{\rm De}^2 - 1}\left[\frac{1}{2} + 2t_D\right] - \left[\frac{3R_{\rm De}^4 - 4R_{\rm De}^4 \ln R_{\rm De} - 2R_{\rm De}^2 - 1}{4(R_{\rm De}^2 - 1)^2}\right] + \sum_{n=1}^{\infty} \frac{2J_1^2(\lambda_n R_{\rm De}) e^{-\lambda_n^2 t_D}}{\lambda_n^2 [J_1^2(\lambda_n R_{\rm De}) - J_1^2(\lambda_n)]}. \qquad (6.4.5)$$

In most cases of practical interest, $R_{\rm De} \gg 1$, and Eq. (6.4.5) can effectively be written as

$$\Delta P_{\rm Dw}(t_D) = \frac{2t_D}{R_{\rm De}^2} + \ln R_{\rm De} - \frac{3}{4} + \sum_{n=1}^{\infty} \frac{2J_1^2(\lambda_n R_{\rm De}) e^{-\lambda_n^2 t_D}}{\lambda_n^2 [J_1^2(\lambda_n R_{\rm De}) - J_1^2(\lambda_n)]}. \qquad (6.4.6)$$

There are three important time regimes in this problem:

(a) A regime in which t_D (henceforth denoted by $t_{\rm Dw}$) is sufficiently large that the logarithmic approximation to the line source solution is valid, but for which the leading edge of the pressure pulse has not yet reached the outer boundary of the reservoir. From Eq. (2.4.7) and Figure 2.3, we find that this regime is defined by

$$25 < t_{\rm Dw} < 0.1R_{\rm De}^2. \tag{6.4.7}$$

As we would expect, Eq. (6.4.5) reduces to the line source solution, Eq. (2.4.10), during this time regime, although this is not obvious from Eq. (6.4.5), or easy to prove.

(b) A second regime that is "late" enough that the presence of the closed outer boundary is felt at the well, but still early enough that the exponential terms in Eq. (6.4.6) have not yet died out. The start of this regime is given by the upper bound in Eq. (6.4.7); it ends when $t_{\rm Dw} \approx 0.3R_{\rm De}^2$ (see Problem 6.1). This regime is therefore defined by

$$0.1R_{\rm De}^2 < t_{\rm Dw} < 0.3R_{\rm De}^2. \tag{6.4.8}$$

In this regime we must use the entire series in Eq. (6.4.6) to calculate the wellbore pressure, and the $\Delta P_{\rm Dw}(t_{\rm Dw})$ curve has no simple description.

(c) A third regime that is late enough such that the exponential terms in Eq. (6.4.6) have effectively *died out.* This regime is defined by

$$t_{\rm Dw} > 0.3R_{\rm De}^2. \tag{6.4.9}$$

In this regime, the wellbore drawdown is given by

$$\Delta P_{\rm Dw} = \frac{2t_{\rm Dw}}{R_{\rm De}^2} + \ln R_{\rm De} - \frac{3}{4}. \tag{6.4.10}$$

Note: These regimes have been given various names in the petroleum engineering literature, such as early-transient, intermediate, late-transient, pseudo-steady-state, semi-steady-state, steady-state, etc. There is no consistent usage of these terms, and many of them are mathematically incorrect. We will refer to the first regime as the "infinite reservoir" regime, the second as the "transition regime", and the last as the "finite reservoir" regime.

An important feature of the finite reservoir regime is that the pressure in the well *declines linearly with time.* The rate of pressure decline can be used to find the drainage area of the well, as follows. First, rewrite Eq. (6.4.10) in terms of the actual variables, rather than the dimensionless variables, so that the pressure at the well is given by

$$P_w(t) = P_i - \frac{Q\mu}{2\pi kH}\left[\frac{2kt}{\phi\mu c R_e^2} + \ln\left(\frac{R_e}{R_w}\right) - \frac{3}{4}\right]. \tag{6.4.11}$$

Now take the derivative of P_w with respect to t:

$$\frac{dP_w}{dt} = -\frac{Q\mu}{2\pi kH}\cdot\frac{2k}{\phi\mu c R_e^2} = \frac{-Q}{\pi R_e^2 H\phi c}. \tag{6.4.12}$$

The rate of change of the late-time well pressure can therefore be used to find the radius of the drainage area, or, equivalently, the drainage area A, i.e.

$$R_e = \left[\frac{-Q}{\pi H\phi c(dP_w/dt)}\right]^{1/2}, \tag{6.4.13}$$

$$A = \pi R_e^2 = \frac{-Q}{\phi c H(dP_w/dt)}. \tag{6.4.14}$$

Equation (6.4.12) can also be "derived" in the following simpler way, by doing a mass balance on the oil in the reservoir:

(a) Imagine that the entire reservoir is at a uniform pressure, P (i.e. a zero-dimensional model, so to speak).
(b) Let $M = \rho V\phi$ be the total amount of oil in the reservoir.
(c) Following the line of reasoning of Section 1.6, the change on the mass of oil contained in the reservoir is related to the change in pressure by

$$\frac{dM}{dt} = \rho V\phi c_t\frac{dP}{dt}. \tag{6.4.15}$$

(d) $-dM/dt$ is the *mass* flow rate *out* of the reservoir, and Q is the volumetric flow rate, so $dM/dt = -\rho Q$, in which case

$$-Q = V\phi c \frac{dP}{dt}. \tag{6.4.16}$$

(e) The volume of the reservoir is $V = \pi R_e^2 H$, so Eq. (6.4.16) is equivalent to Eq. (6.4.12)!

The dimensionless wellbore pressure for a well located at the centre of a bounded circular reservoir is plotted from Eq. (6.4.6) in Figure 6.6, for various values of the dimensionless reservoir size, $R_{\text{De}} = R_e/R_w$. Also shown is the wellbore pressure for the case of constant flow rate into the borehole and constant pressure at the outer boundary, from Eq. (6.3.10). Note that:

- In accordance with Eq. (6.4.7), the semi-log straight line begins at about $t_{\text{Dw}} = 25$.
- In accordance with Eq. (6.4.8), the "infinite reservoir" regime ends when $t_{\text{Dw}} = 0.1R_{\text{De}}^2$.

For example, when $R_{\text{De}} = 1{,}000$, the curve begins to deviate from the semi-log straight line at about $t_{\text{Dw}} = 0.1R_{\text{De}}^2 = 1 \times 10^5$.

6.5. Non-circular Drainage Regions

If the drainage region is not circular, and/or the well is not located at the centre, then the problem treated in Section 6.4 is more difficult to solve. If the drainage region is bounded by a polygon, as would occur if the reservoir was bounded by a set of linear faults, then the problem can be solved by superposing solutions from appropriately located *image wells*, as in Section 4.3. For example, an infinite array of production image wells arranged on a square lattice will create a square reservoir that is bounded by four no-flow boundaries. Examples of this type of analysis can be found in the monograph *Well Testing in Heterogeneous Formations* by Streltsova (1988).

Without going into the details of this analysis, the following can be said about a well producing from within a non-circular drainage region. There is an "infinite reservoir" regime, during which the pressure pulse has not yet reached any portion of the outer boundary. The duration of this regime can be calculated using Eq. (1.6.11), if the geometry of the drainage area and the location of the well are known. During this regime,

$$\Delta P_{\mathrm{Dw}} = \frac{1}{2}[\ln(t_{\mathrm{Dw}}) + 0.80907]. \tag{6.5.1}$$

In this regime, the overall shape of the reservoir is irrelevant, and the drawdown is the same as that which would occur in an unbounded reservoir.

At sufficiently long times, the mass-balance argument presented in Section 6.4 must hold, and so the *rate* at which the wellbore drawdown changes must be given by Eq. (6.4.12), with R_e^2 replaced by A/π. Hence, an equation of the same form as Eq. (6.4.10) must hold for the drawdown, except that the constant term may be different. The duration of this regime is roughly given by Eq. (6.4.9), again with R_e^2 replaced by A/π, although the start of this regime is delayed if the well is located off-centre of the drainage region by an appreciable amount. During this regime, the dimensionless wellbore drawdown is traditionally written in the following form:

$$\Delta P_{\mathrm{Dw}} = 2\pi t_{\mathrm{DA}} + \frac{1}{2}\ln\left(\frac{4A}{\gamma R_w^2 C_A}\right), \tag{6.5.2}$$

where $\gamma = 1.781$, C_A is a dimensionless constant known as the "Dietz shape factor" (Dietz, 1965), and t_{DA} is a dimensionless time based on the drainage area:

$$t_{\mathrm{DA}} = \frac{kt}{\phi\mu c A}. \tag{6.5.3}$$

Comparison of Eq. (6.5.2) with Eq. (6.4.10) reveals that, for a well at the centre of a circular reservoir,

$$C_A = \frac{4\pi}{\gamma e^{-3/2}} = 31.62. \tag{6.5.4}$$

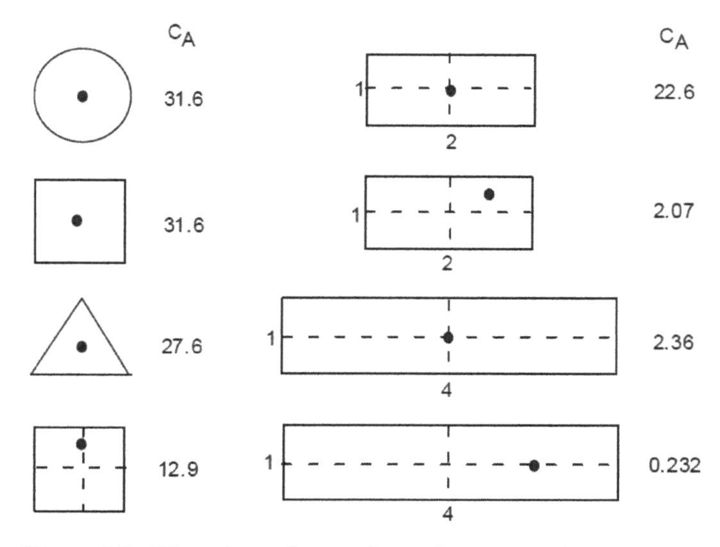

Figure 6.7. Dietz shape factors for various reservoir geometries.

The shape factor tends to decrease as the drainage area becomes more non-circular, or as the location of the well becomes more off-centre. Shape factors for numerous geometries can be found on p. 111 of Matthews and Russell (1967); a few cases are shown in Figure 6.7.

Problems for Chapter 6

Problem 6.1. Starting with the expression for the pressure drawdown, Eq. (6.4.6), show that the transition regime for a closed circular reservoir with constant production rate ends when $t \approx 0.3\phi\mu c R_e^2/k$.

Hints:

(a) $e^{-x} \approx 0$ when $x > 4$, so all terms in the series will be negligible when $\lambda_n^2 t_D > 4$ for all n.
(b) When R_{De} is large, the *first* eigenvalue, λ_1, defined as the smallest value of λ that satisfies Eq. (6.4.4), is *very* small. This

fact should help you to estimate the value of λ_1 as a function of R_{De}, by making the reasonable assumption that λ_1 is inversely proportional to R_{De}.

(c) Make use of Eqs. (6.2.37), (6.2.39), (6.2.48) and Figure 6.3.

Problem 6.2. Starting with Eq. (6.4.2), calculate the average pressure in the reservoir during the finite reservoir regime, $t_{\mathrm{Dw}} \equiv t_D > 0.3R_{\mathrm{De}}^2$, as a function of time. Is your result consistent with the mass-balance that is embodied in Eq. (6.4.16)?

Problem 6.3. Starting with Eq. (6.3.2), calculate the average pressure in a circular reservoir with a constant-pressure outer boundary, during the late-time regime in which all of the exponential terms have died out. Use this result to find an equation for the well productivity, which relates the production rate to the difference between the average reservoir pressure and the pressure at the well.

Chapter 7

Laplace Transform Methods in Reservoir Engineering

The most widely used method for solving the pressure diffusion equation, for different reservoir geometries and different boundary conditions, is the method of Laplace transforms. Using this approach, the partial differential diffusion equation is transformed into an ordinary differential equation, in the "Laplace domain", which in general is straightforward to solve. The solution in the Laplace domain is then transformed back into the "time domain", using well-known analytical or numerical algorithms.

In this chapter, we will introduce the Laplace transform method, present many of the useful properties of the Laplace transform that facilitate its use in solving the pressure diffusion equation, and then give an example of its use in solving an important reservoir engineering problem: flow to well containing a vertical hydraulic fracture. Finally, the concept of convolution, introduced in Chapter 3, will be shown to have a particularly simple and elegant form when expressed in the Laplace domain.

7.1. Introduction to Laplace Transforms

The diffusion equation for fluid flow in a porous medium, in the linearised form that was derived in Chapter 1 and used thus far in these notes, is a *linear* partial differential equation (PDE). As such, many of the classical methods of applied mathematics can be used to solve it. In Section 2.1, we used the Boltzmann transformation to solve the problem of a single, fully penetrating well in an infinite, homogeneous reservoir. We then learned in Section 6.1 that this method is only applicable to problems in unbounded reservoirs.

In Chapter 6, we used the method of eigenfunction expansions to solve problems involving a well in a circular reservoir. Although the method of eigenfunction expansions is the most widely used method in applied mathematics, and was used by Muskat (1937) in his monumental book *The Flow of Homogeneous Fluids Through Porous Media*, since the 1940s, reservoir engineers have tended to solve the pressure diffusion equation using Laplace transforms.

The method of Laplace transforms allows us to transform linear PDEs, which are generally difficult to solve, into linear ordinary differential equations (ODEs), which can be solved in essentially all cases. Specifically, the function $P(R,t)$, which satisfies a PDE, is formally transformed into a different function $\widehat{P}(R,s)$, which satisfies an ODE, where s is a Laplace variable that plays the role of a parameter (since no derivatives are taken with respect to s). This ODE is then solved to find $\widehat{P}(R,s)$. The most difficult part of the method of Laplace transforms is the process of "inverting" the function from the "Laplace domain" back into the "time domain", to recover $P(R,t)$. This last step can be achieved either by contour integration in the complex plane, or by numerical integration along the real axis, as will be explained below.

In this section, we present the definition of the Laplace transform, along with a few of its more important properties. A fuller treatment can be found in essentially any textbook on applied mathematics. Three books that are devoted exclusively to Laplace (and related) transforms are as follows:

(1) R. V. Churchill (1958) *Operational Mathematics*, McGraw-Hill.
(2) H. Carslaw and J. C. Jaeger (1949) *Operational Methods in Applied Mathematics*, Oxford University Press.
(3) C. J. Tranter (1971) *Integral Transforms in Mathematical Physics*, Chapman & Hall.

Definition: If we have a function $f(t)$, its Laplace transform is defined by

$$L\{f(t)\} \equiv \widehat{f}(s) \equiv \int_0^\infty f(t)e^{-\mathrm{st}}dt. \tag{7.1.1}$$

Both notations, $L\{f(t)\}$ and $\widehat{f}(s)$, are useful, depending on the context. In the general theory of Laplace transforms, the parameter s must be regarded as a complex variable, but for our purposes we can usually think of it as a real number.

If we have a function of two variables, such as $P(R,t)$, we can define its Laplace transform in the same manner:

$$\widehat{P}(R,s) \equiv \int_0^\infty P(R,t)e^{-\mathrm{st}}dt. \tag{7.1.2}$$

Note that the Laplace transform is taken with respect to the time variable, not the spatial variable. To simplify the notation, we will usually suppress the R variable when discussing the general theory.

The Laplace transform operator L is a *linear operator*, because it follows directly from definition (7.1.1) that

$$L\{cf(t)\} = \int_0^\infty cf(t)e^{-\mathrm{st}}dt = c\int_0^\infty f(t)e^{-\mathrm{st}}dt = cL\{f(t)\}, \tag{7.1.3}$$

$$\begin{aligned} L\{f_1(t) + f_2(t)\} &= \int_0^\infty [f_1(t) + f_2(t)]e^{-\mathrm{st}}dt \\ &= \int_0^\infty f_1(t)e^{-\mathrm{st}}dt + \int_0^\infty f_2(t)e^{-\mathrm{st}}dt \\ &= L\{f_1(t)\} + L\{f_2(t)\}. \end{aligned} \tag{7.1.4}$$

The most important property of the Laplace transform is that differentiation in the time domain essentially corresponds to multiplication by s in the Laplace domain. To prove this, consider the Laplace transform of the time derivative of $f(t)$:

$$L\{f'(t)\} \equiv \int_0^\infty f'(t)e^{-\mathrm{st}}dt. \tag{7.1.5}$$

Now recall the general formula for integration by parts:

$$\int_0^\infty f'(t)g(t)dt = f(t)g(t)]_0^\infty - \int_0^\infty f(t)g'(t)dt. \tag{7.1.6}$$

If we apply formula (7.1.6), with $g(t) = e^{-\mathrm{st}}$, we find

$$L\{f'(t)\} = f(t)e^{-\mathrm{st}}]_0^\infty + s\int_0^\infty f(t)e^{-\mathrm{st}}dt,$$

$$= f(\infty)e^{-s\cdot\infty} - f(0)e^{-s\cdot 0} + sL\{f(t)\}$$

$$\text{i.e.}\quad L\{f'(t)\} = sL\{f(t)\} - f(0). \tag{7.1.7}$$

Hence, the Laplace transform of $f'(t)$ is equal to the Laplace transform of $f(t)$ multiplied by s, *minus* the initial value of $f(t)$.

Equation (7.1.7) illustrates two important properties of Laplace transforms:

(a) Differentiation in the time domain corresponds (essentially) to multiplication by s in the Laplace domain.
(b) The initial conditions become incorporated directly into the governing equation. This is unlike the situation in "time domain" methods, where initial conditions must be considered separately after we have found the general solution to the differential equation.

Note: Since we usually work with pressure "drawdowns", which are *defined* so as to be zero when $t = 0$, the $f(0)$ term will usually drop out of our calculations.

As differentiation with respect to time corresponds to *multiplication* by s in the Laplace domain, we might expect, correctly, that integration with respect to time corresponds to *division* by s in the Laplace domain. To prove this, let $F(t)$ be the time integral of $f(t)$, i.e.

$$F(t) \equiv \int_0^t f(\tau)d\tau, \tag{7.1.8}$$

where τ is a dummy variable of integration, used to avoid confusion with the specific value of t that appears at the upper limit of integration. By definition, then,

$$F'(t) = f(t), \quad F(0) \equiv \int_0^0 f(\tau)d\tau = 0. \tag{7.1.9}$$

Now, using property (7.1.7),

$$L\{F'(t)\} = sL\{F(t)\} - F(0) = sL\{F(t)\} = sL\left\{\int_0^t f(\tau)d\tau\right\},$$

which can be rearranged to give

$$L\left\{\int_0^t f(\tau)d\tau\right\} = \frac{1}{s}L\{F'(t)\} = \frac{1}{s}L\{f(t)\} = \frac{1}{s}\widehat{f}(s), \tag{7.1.10}$$

which shows that, in order to find the Laplace transform of the *integral* of $f(t)$, we merely have to take the Laplace transform of $f(t)$, and divide it by s. Note that since the "initial condition" of a definite integral is zero, no initial condition term appears in this relation.

Another useful property of Laplace transforms is the "time-shift" property. Consider a function $f(t)$, and the time-shifted function $f(t - t_0)$, defined by the graph in Figure 7.1.

Note that $f(t - t_0)$ is defined to be 0 if $t < t_0$, which is consistent with the fact that, when using Laplace transforms, we always consider *all* functions $f(t)$ to equal zero when $t < 0$. (This assumption is consistent with the physical fact that the pressure drawdown will be zero before production begins!)

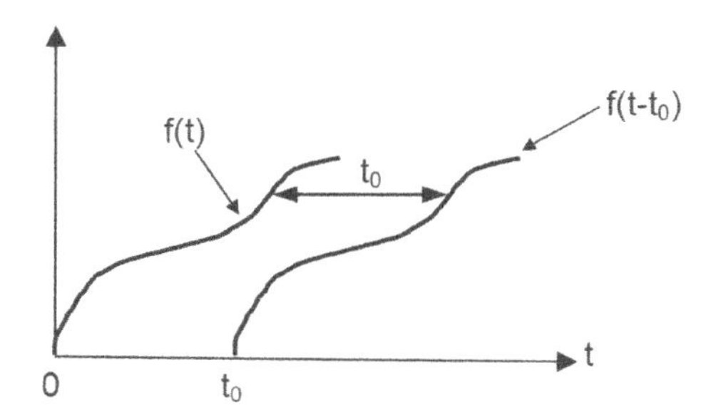

Figure 7.1. A function, $f(t)$, and its time-shifted variant, $f(t - t_0)$.

The Laplace transform of $f(t-t_0)$ can be found directly from the Laplace transform of $f(t)$, as follows. First, by definition,

$$L\{f(t-t_0)\} = \int_0^\infty f(t-t_0)e^{-st}dt. \quad (7.1.11)$$

Now do a change of variables, letting $t-t_0 = \tau$, in which case $dt = d\tau$. The limits of integration change, according to

$$\tau = -t_0 \quad \text{when } t = 0, \qquad \tau = \infty \quad \text{when } t = \infty. \quad (7.1.12)$$

Hence,

$$L\{f(t-t_0)\} = \int_{-t_0}^\infty f(\tau)e^{-s(\tau+t_0)}d\tau. \quad (7.1.13)$$

But $f(t) = 0$ when $t < 0$, so

$$L\{f(t-t_0)\} = \int_0^\infty f(\tau)e^{-s(\tau+t_0)}d\tau = \int_0^\infty f(\tau)e^{-s\tau}e^{-st_0}d\tau$$

$$= e^{-st_0}\int_0^\infty f(\tau)e^{-s\tau}d\tau = e^{-st_0}\widehat{f}(s). \quad (7.1.14)$$

So, we see that "delaying" a function $f(t)$ by an amount of time t_0 corresponds to multiplying its Laplace transform by e^{-st_0}.

If we take a function $f(t)$, and "damp it out" by multiplying it by e^{-at}, this is equivalent to replacing s with $s+a$ in the Laplace transform of $f(t)$. The proof is as follows:

$$L\{e^{-at}f(t)\} = \int_0^\infty f(t)e^{-at}e^{-st}dt$$

$$= \int_0^\infty f(t)e^{-(s+a)t}dt = \widehat{f}(s+a). \quad (7.1.15)$$

Finally, consider a function $f(t)$, and the "stretched" function $f(at)$. The Laplace transform of $f(at)$ is related to the Laplace transform of $f(t)$ as follows (see Problem 7.2):

$$L\{f(at)\} = \frac{1}{a}\widehat{f}(s/a). \quad (7.1.16)$$

If we think of a as a frequency, as in the function $\sin(at)$, Eq. (7.1.16) shows that "speeding up" a function by a factor a is somewhat equivalent to "slowing down" its Laplace transform by the factor $1/a$.

The usefulness of the rules presented above is that they minimise the number of times we actually have to calculate a Laplace transform using the basic definition (7.1.1). By knowing the Laplace transforms of a few functions, we can use these rules to calculate the Laplace transforms of many other functions. For example, first note that if $f(t) = 1$, then

$$L\{f(t)\} = L\{1\} = \int_0^\infty 1e^{-\mathrm{st}}dt = \frac{-1}{s}e^{-\mathrm{st}}]_0^\infty = \frac{1}{s}. \tag{7.1.17}$$

Aside: The function $f(t) = 1$ is very well behaved mathematically, but its Laplace transform is not, since $\widehat{f}(s) \to \infty$ when $s = 0$. Points where $\widehat{f}(s)$ becomes infinite, or is otherwise non-analytic (i.e. cannot be represented by a Taylor series), are known as *singularities.* The singularities of $\widehat{f}(s)$ are actually the keys to "inverting" $\widehat{f}(s)$ in order to find the actual function, $f(t)$. However, we will not pursue these ideas very far, because they require knowledge of complex variable theory.

Now suppose that we want to calculate $L\{t\}$. We could use Eq. (7.1.1), but it is easier to see from Eq. (7.1.10) that, since t is the integral of 1, then $L\{t\}$ can be found by dividing $L\{1\}$ by s, i.e.

$$L\{t\} = L\left\{\int_0^t 1\,d\tau\right\} = \frac{1}{s}L\{1\} = \frac{1}{s^2}. \tag{7.1.18}$$

Repeated application of Eq. (7.1.10) leads to a general expression for the Laplace transform of t^n, where n is any non-negative integer (see Problem 7.3):

$$L\{t^n\} = \frac{n!}{s^{n+1}}. \tag{7.1.19}$$

Another Laplace transform that appears often when solving diffusion problems is $L\{t^{-1/2}\}$:

$$L\{t^{-1/2}\} = \int_0^\infty t^{-1/2}e^{-\mathrm{st}}dt. \tag{7.1.20}$$

To evaluate this integral, first make the change of variables $st = u$, in which case $dt = du/s$, and the integral becomes

$$L\{t^{-1/2}\} = \int_0^\infty (u/s)^{-1/2} e^{-u} \frac{du}{s} = \frac{1}{\sqrt{s}} \int_0^\infty u^{-1/2} e^{-u} du. \quad (7.1.21)$$

Now put $u = m^2$, in which case $du = 2mdm = 2\sqrt{u}dm$, and so

$$L\{t^{-1/2}\} = \frac{2}{\sqrt{s}} \int_0^\infty e^{-m^2} dm = \frac{2}{\sqrt{s}} \frac{\sqrt{\pi}}{2} = \sqrt{\frac{\pi}{s}}, \quad (7.1.22)$$

which can also be expressed as

$$L^{-1}\{\pi^{1/2} s^{-1/2}\} = t^{-1/2}. \quad (7.1.23)$$

By applying the rule that division by s in the Laplace domain corresponds to integration with respect to t in the time domain, we can show, starting with Eq. (7.1.23), that

$$L^{-1}\{\pi^{1/2} s^{-3/2}\} = \int_0^t \tau^{-1/2} d\tau = 2t^{1/2}$$

$$\rightarrow L\{t^{1/2}\} = \frac{\sqrt{\pi}}{2s^{3/2}}. \quad (7.1.24)$$

This process can be repeated to give, for $n = 1, 2, \ldots,$

$$L\{t^{n-(1/2)}\} = \frac{1 \cdot 3 \cdot 5 \cdot \ldots \cdot (2n-1)\sqrt{\pi}}{2^n s^{n+(1/2)}}. \quad (7.1.25)$$

For example, for $n = 3$, Eq. (7.1.25) shows that $L\{t^{5/2}\} = 15\sqrt{\pi}/8s^{7/2}$.

The most difficult aspect of using Laplace transforms is "inverting" the Laplace transform $\widehat{f}(s)$ to find the function $f(t)$. The difficulty lies in the fact that, even though we, as petroleum engineers, are ultimately interested only in real-valued functions, the standard inversion procedure involves integration *in the complex plane*. The inversion formula, which allows us to recover $f(t)$ from $\widehat{f}(s)$, is as

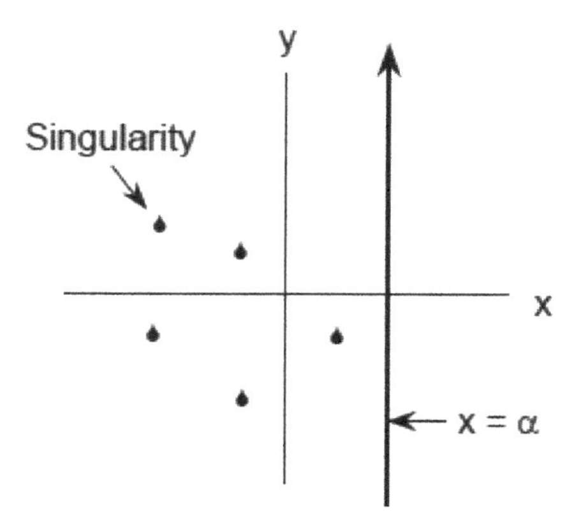

Figure 7.2. Path of integration for the Laplace transform inversion integral given in Eq. (7.1.26).

follows:

$$f(t) = \frac{1}{2\pi i} \int_{\alpha - i\infty}^{\alpha + i\infty} \widehat{f}(s) e^{st} ds, \tag{7.1.26}$$

where $i = \sqrt{-1}$ is the imaginary unit number, and the integration takes place along any vertical line in the complex plane that lies to the *right* of all the singularities of $\widehat{f}(s)$; see Figure 7.2, in which $s = x + iy$. The proof of Eq. (7.1.26), which is not simple, can be found in the aforementioned books by Carslaw and Jaeger (1949), or Churchill (1958).

One additional fact about Laplace transforms that is needed in order to use this method to solve the pressure diffusion equation is the following rule:

- If $P(R, t)$ is a function of R and t, and $\widehat{P}(R, s)$ is its Laplace transform, as defined in Eq. (7.1.2), then the Laplace transform of the partial derivative of P with respect to R is equal to the partial derivative, with respect to R, of the Laplace transform of P, i.e.

$$L\left\{\frac{dP}{dR}\right\} = \frac{d}{dR}[L\{P(R, t)\}] = \frac{d\widehat{P}(R, s)}{dR}. \tag{7.1.27}$$

The proof of this rule follows directly from Leibnitz's theorem for differentiating an integral with respect to a parameter that appears in the integrand:

$$L\left\{\frac{dP(R,t)}{dR}\right\} = \int_0^\infty \frac{dP(R,t)}{dR} e^{-st} dt$$
$$= \frac{d}{dR}\int_0^\infty P(R,t)e^{-st}dt = \frac{d\widehat{P}(R,s)}{dR}. \quad (7.1.28)$$

In the second term we first differentiate with respect to R, and then integrate with respect to t, whereas in the third term we first integrate with respect to t, and then differentiate with respect to R. Leibnitz's theorem tells us that the order in which we carry out these two operations does not matter.

7.2. Flow to a Hydraulically Fractured Well

In general, exact inversion (as opposed to numerical inversion, which will be discussed in Section 7.4) of a LT requires knowledge of complex variable theory. One of the few important flow problems that we can solve with LTs without knowledge of complex integration is the one-dimensional (1D) diffusion equation in a semi-infinite region.

The 1D pressure diffusion problem is relevant to flow to a "hydraulically fractured" well, for example. In low permeability reservoirs, fractures are often introduced into the rock by pumping fluid into the borehole at high pressure; see the geomechanics module of this MSc course. These "hydraulic fractures" then provide very conductive paths for the oil to reach the wellbore. Oil first flows to the fracture, and then through the fracture to the well.

Imagine that the thickness of the reservoir is H, and that the hydraulic fracture extends out from the vertical borehole by a distance L in each direction, as shown in Figure 7.3. Initially, fluid flows straight to the nearest part of the fracture, after which it travels through the hydraulic fracture to the wellbore. If the length L is large, or equivalently, at early times, we can model this process as uniform, 1D, horizontal flow.

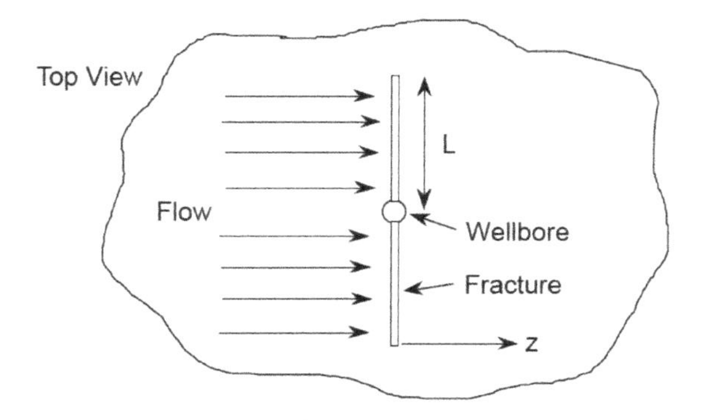

Figure 7.3. Schematic diagram of a 1D flow to a hydraulically fractured well.

The governing equation for this problem is the 1D pressure diffusion equation in Cartesian co-ordinates, Eq. (1.6.8):

$$\text{PDE:} \quad \frac{dP}{dt} = D\frac{d^2P}{dz^2}, \tag{7.2.1}$$

where z is the coordinate normal to the fracture plane, and D is the hydraulic diffusivity, $k/\phi\mu c_t$.

If the total flow rate into the well is Q, and this flow is distributed uniformly over an area of $4LH$(two fractures with two faces, each of area LH), then the initial and boundary conditions are

$$\text{IC:} \quad P(z, t = 0) = P_i, \tag{7.2.2}$$

$$\text{far-field BC:} \quad P(z \to \infty, t) = P_i, \tag{7.2.3}$$

$$\text{fracture BC:} \quad \frac{dP}{dz}(z = 0, t) = \frac{\mu Q}{4kLH}. \tag{7.2.4}$$

To solve this problem, we first define the LT of $P(z, t)$:

$$\widehat{P}(z, s) \equiv \int_0^\infty P(z, t)e^{-\text{st}}dt. \tag{7.2.5}$$

Next, we take the LT of *both sides* of Eq. (7.2.1). Using Eq. (7.1.7), and Eq. (7.2.2), the LHS of Eq. (7.2.1) is transformed as follows:

$$L\left\{\frac{dP}{dt}\right\} = sL\{P(z,t)\} - P(z,t=0) = s\,\widehat{P}(z,s) - P_i. \tag{7.2.6}$$

Applying rule (7.1.27) twice, the LT of the RHS of Eq. (7.2.1) is

$$L\left\{D\frac{d^2P}{dz^2}\right\} = D\frac{d^2}{dz^2}[L\{P(z,t)\}] = D\frac{d^2\,\widehat{P}(z,s)}{dz^2}. \tag{7.2.7}$$

So, the transformed representation of Eq. (7.2.1) is

$$\text{ODE} : D\frac{d^2\,\widehat{P}(z,s)}{dz^2} - s\,\widehat{P}(z,s) = -P_i. \tag{7.2.8}$$

Although $\widehat{P}(z,s)$ is a function of two variables, z and s, no derivatives of $\widehat{P}(z,s)$ with respect to s appear in Eq. (7.2.8). So, s actually appears in Eq. (7.2.8) as a *parameter*, not a *variable*. Consequently, Eq. (7.2.8) is an ODE, rather than a PDE.

The initial condition is already incorporated into Eq. (7.2.8). However, we must now take the LTs of the two BCs, Eqs. (7.2.3) and (7.2.4):

Far-field BC:

$$L\{P(z=\infty,t)\} = \widehat{P}(z=\infty,s) = L\{P_i\} = \frac{P_i}{s}, \tag{7.2.9}$$

Fracture BC:

$$L\left\{\frac{dP}{dz}(z=0,t)\right\} = \frac{d\widehat{P}}{dz}(z=0,s) = L\left\{\frac{\mu Q}{4kLH}\right\} = \frac{\mu Q}{4kLHs}. \tag{7.2.10}$$

Next, we solve the problem in the Laplace domain. The general solution to Eq. (7.2.8) is

$$\widehat{P}(z,s) = Ae^{z\sqrt{s/D}} + Be^{-z\sqrt{s/D}} + \frac{P_i}{s}, \tag{7.2.11}$$

where A and B are arbitrary constants. Applying the far-field boundary condition, Eq. (7.2.9), to the general solution (7.2.11), shows us that A must be zero. Applying the fracture boundary condition, Eq. (7.2.10), to the general solution (7.2.11), implies that

$$\frac{d\widehat{P}}{dz}(z=0,s) = -B\sqrt{s/D} = \frac{\mu Q}{4kLHs},$$
$$\rightarrow B = \frac{-\mu Q}{4kLHs\sqrt{s/D}}. \tag{7.2.12}$$

So, we have now found A and B, and the solution to this problem *in the Laplace domain* is

$$\widehat{P}(z,s) = \frac{P_i}{s} - \frac{\mu Q\sqrt{D}}{4kLHs^{3/2}}e^{-z\sqrt{s/D}}. \tag{7.2.13}$$

Finally, we must invert $\widehat{P}(z,s)$ to find $P(z,t)$. To keep the analysis simple and brief, we will only invert for the pressure in the fracture, $P(z=0,t)$, which is actually the pressure that we are most interested in. If the permeability of the fracture is much greater than that of the reservoir (which is the idea behind hydraulic fracturing!), then there will be very little pressure drop in the fracture itself, and the pressure in the fracture will be equal to the pressure in the wellbore. So, in the fracture,

$$\widehat{P}(z=0,s) = \frac{P_i}{s} - \frac{\mu Q\sqrt{D}}{4kLHs^{3/2}}. \tag{7.2.14}$$

Note: This illustrates another advantage of LTs: we can usually find the wellbore pressure *without* having first to find the pressure at every point in the reservoir. With most other mathematical methods, you *must* find $P(z,t)$ first as a complete function, and *then* set $z=0$.

We now "invert" to find $P(z=0,t)$, using Eqs. (7.1.17) and (7.1.24), to find (Stewart, 2011, p. 478):

$$P(z=0,t) = L^{-1}\{\widehat{P}(z=0,s)\} = L^{-1}\left\{\frac{P_i}{s} - \frac{\mu Q\sqrt{D}}{4kLHs^{3/2}}\right\}$$
$$= P_i L^{-1}\left\{\frac{1}{s}\right\} - \frac{\mu Q\sqrt{D}}{4kLH}L^{-1}\left\{\frac{1}{s^{3/2}}\right\}$$
$$= P_i - \frac{\mu Q}{2kLH}\sqrt{\frac{Dt}{\pi}}. \tag{7.2.15}$$

Equation (7.2.15) shows that at early times, the drawdown in a fractured well will increase in proportion to $t^{1/2}$.

7.3. Convolution Principle in the Laplace Domain

The convolution integral that appeared in Eq. (3.4.6) is a particular case of a more general mathematical operation that can be defined by

$$f * g \equiv \int_0^t f(\tau)g(t-\tau)d\tau, \tag{7.3.1}$$

which is known as the *convolution* of the two functions f and g.

If we let $t-\tau = x$ in Eq. (7.3.1), then $\tau = t - x$ and $d\tau = -dx$, and the limits of integration are transformed into $x = t$ and $x = 0$. Hence,

$$f * g = -\int_t^0 f(t-x)g(x)dx = \int_0^t f(t-x)g(x)dx \equiv g * f, \tag{7.3.2}$$

which shows that $f * g = g * f$, for any two functions f and g.

Now, let us look at the LT of $f * g$:

$$L\{f * g\} = \int_{t=0}^{t=\infty} \left[\int_{\tau=0}^{\tau=t} f(\tau)g(t-\tau)d\tau\right] e^{-\mathrm{st}}dt. \tag{7.3.3}$$

The region of integration covers the first octant of the $\{t,\tau\}$ plane, as can be seen from Figure 7.4. The thick line on the left runs from $\tau = 0$ to $\tau = t$, for fixed t, which corresponds to the inner integral in Eq. (7.3.3). If we sweep this line across from left to right, then t

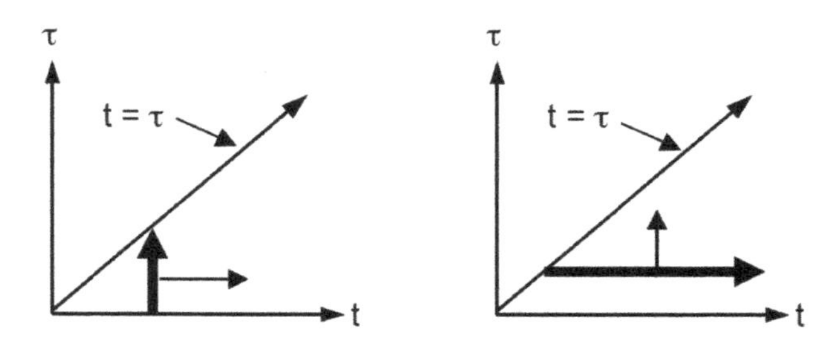

Figure 7.4. Two paths of integration for Eq. (7.3.3).

will cover the range from $t = 0$ to $t = \infty$, which corresponds to the outer integral.

But this octant can also be covered by first letting t run from $t = \tau$, out to $t = \infty$, for fixed τ (see the thick line on right), and then letting τ run from $\tau = 0$ to $\tau = \infty$. Hence, we can also write the convolution integral as

$$L\{f * g\} = \int_{\tau=0}^{\tau=\infty} \left[\int_{t=\tau}^{t=\infty} g(t-\tau)e^{-\mathrm{st}}dt \right] f(\tau)d\tau. \tag{7.3.4}$$

If we now let $t - \tau = x$ in Eq. (7.3.4), then $t = \tau + x$, $dt = dx$, and the limits of integration of the inner integral are transformed into $x = 0$ and $x = \infty$, and so

$$\begin{aligned} L\{f * g\} &= \int_{\tau=0}^{\tau=\infty} \left[\int_{x=0}^{x=\infty} g(x)e^{-s(\tau+x)}dx \right] f(\tau)d\tau \\ &= \left[\int_{\tau=0}^{\tau=\infty} f(\tau)e^{-s\tau}d\tau \right] \left[\int_{x=0}^{x=\infty} g(x)e^{-\mathrm{sx}}dx \right] = \widehat{f}(s)\,\widehat{g}(s). \end{aligned} \tag{7.3.5}$$

We have therefore proven that convolution of two functions in the time domain corresponds to multiplication of the their LTs. This fact is extremely useful, because:

(a) It implies that, if we can break up a difficult LT into the product of two simpler transforms whose inverses are known, then we can find the complete inverse function, in the time domain, by merely doing a convolution integral.

(b) More importantly, it implies that, for any given reservoir geometry, if we can solve the diffusion equation for the case of a constant production rate, then the solution for an arbitrary production schedule can be found by convolution.

Actually, we already knew this from Section 3.4. However, the concept is more general, since, for example, we can also use convolution to find the solution for the case of *varying* wellbottom pressure from the solution for the case of *constant* wellbottom pressure.

As an example of the use of convolution in the Laplace domain, consider the hydraulic fracture problem of Section 7.2, but with an arbitrary *time-dependent* flow rate into the fracture, $Q(t)$. The only change in the formulation of the problem will be in the fracture boundary condition, Eq. (7.2.4), which now becomes

$$\frac{dP}{dz}(z=0,t) = \frac{\mu Q(t)}{4kLH}. \tag{7.3.6}$$

If we follow the solution procedure used in Section 7.2, the only change occurs when we take the LT of this boundary condition, in which case $\mu Q/4kLHs$ is replaced with the more general expression $\mu \widehat{Q}(s)/4kLH$. The solution in Laplace space then becomes

$$\widehat{P}(z,s) = \frac{P_i}{s} - \frac{\mu \widehat{Q}(s)\sqrt{D}}{4kLHs^{1/2}} e^{-z\sqrt{s/D}}. \tag{7.3.7}$$

We again restrict our attention to the fracture, where $z = 0$:

$$\widehat{P}(z=0,s) = \frac{P_i}{s} - \frac{\mu \widehat{Q}(s)\sqrt{D}}{4kLHs^{1/2}}. \tag{7.3.8}$$

We now take the inverse LT, and in doing so make use of linearity and convolution:

$$\begin{aligned} P(z=0,t) &= L^{-1}\{\widehat{P}(z=0,s)\} = L^{-1}\left\{\frac{P_i}{s} - \frac{\mu \widehat{Q}(s)\sqrt{D}}{4kLHs^{1/2}}\right\} \\ &= P_i L^{-1}\left\{\frac{1}{s}\right\} - \frac{\mu\sqrt{D}}{4kLH} L^{-1}\left\{\frac{\widehat{Q}(s)}{s^{1/2}}\right\} \\ &= P_i - \frac{\mu\sqrt{D}}{4kLH}\left[L^{-1}\{\widehat{Q}(s)\}\right] * \left[L^{-1}\{s^{-1/2}\}\right]. \end{aligned} \tag{7.3.9}$$

But, $L^{-1}\{\widehat{Q}(s)\} = Q(t)$, by definition, and $L^{-1}\{s^{-1/2}\} = (\pi t)^{-1/2}$ by Eq. (7.1.23), so

$$P(z=0,t) = P_i - \frac{\mu\sqrt{D}}{4kLH}[Q(t)] * [(\pi t)^{-1/2}]$$

$$= P_i - \frac{\mu\sqrt{D/\pi}}{4kLH}\int_0^t \frac{Q(\tau)}{\sqrt{t-\tau}}d\tau$$

$$= P_i - \frac{1}{4LH}\sqrt{\frac{\mu}{\pi k\phi c}}\int_0^t \frac{Q(\tau)}{\sqrt{t-\tau}}d\tau$$

$$= P_i - \frac{1}{A}\sqrt{\frac{\mu}{\pi k\phi c}}\int_0^t \frac{Q(\tau)}{\sqrt{t-\tau}}d\tau$$

$$= P_i - \sqrt{\frac{\mu}{\pi k\phi c}}\int_0^t \frac{q(\tau)}{\sqrt{t-\tau}}d\tau. \qquad (7.3.10)$$

Equation (7.3.10) allows us to find the pressure in the fracture as a function of the time-varying flow rate per unit area, $q(t) = Q(t)/A$, by merely doing an integral in the time domain.

Although it may be difficult to evaluate this integral analytically for some particular production rate $q(t)$, it will always be straightforward to evaluate it numerically. Convolution leads to an integral over time, which is a real variable, and such integrals are generally easier to evaluate than the *complex* inversion integral specified in Eq. (7.1.26).

7.4. Numerical Inversion of Laplace Transforms

The complex integral that allows us to invert the LT of the pressure to obtain the pressure as a function of *time* is often very difficult to evaluate in closed form. Consequently, many algorithms have been devised to carry out this inversion numerically. Their use in petroleum engineering is reviewed in *Fundamental and Applied Pressure Analysis* by Daltaban and Wall (1998).

In petroleum engineering, the most widely used such algorithm is the Stehfest (1968) algorithm. Although the derivation of this algorithm is lengthy, and beyond the scope of these notes, we can gain a rough understanding of it as follows.

In general, any integral can be approximated by a summation whose terms consist of the integrand evaluated at various discrete points, with each functional evaluation multiplied by an appropriate

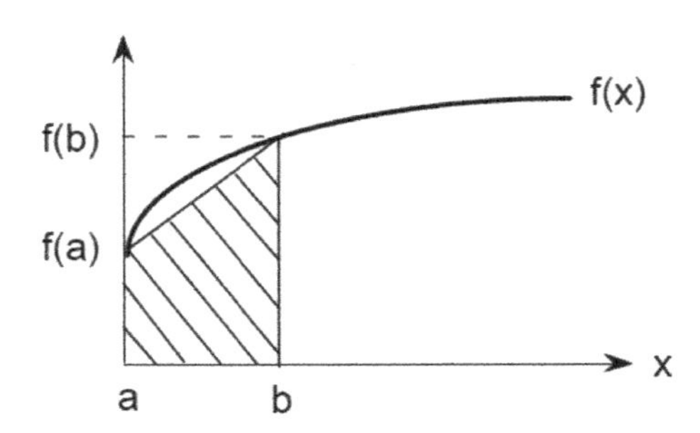

Figure 7.5. Approximation of an integral using the trapezium rule.

weighting factor. For example, consider the numerical approximation of an integral based on the trapezium rule. Imagine a function $f(x)$, which we want to integrate from $x = a$ to $x = b$. Recall that an integral can be interpreted as the area under the graph of the function.

The area under the graph of $f(x)$ can be approximated by the area of the trapezium shown in Figure 7.5, i.e.

$$\int_a^b f(x)dx \approx \left[\frac{f(a) + f(b)}{2}\right](b - a). \tag{7.4.1}$$

This formula approximates the integral by evaluating the integrand at $x = a$ and $x = b$, multiplying the two function values by the weighting factor $(b - a)/2$, and then summing. More accurate numerical approximations of an integral can be made by:

(a) evaluating the integrand at a larger number of points,
(b) carefully choosing the points at which the integrand is evaluated (i.e. they need not be equally spaced),
(c) carefully choosing the weighting factors.

With these ideas in mind, we can think of the Stehfest algorithm as a clever method for approximating the complex inversion integral by a weighted finite sum of function evaluations.

If we have the LT of, say, the wellbore pressure $\widehat{P_w}(s)$, the actual wellbore pressure at time t is given, according to

Stehfest, by

$$P_w(t) = \frac{\ln 2}{t}\sum_{n=1}^{2N} V_n \widehat{P}_w\left(s = \frac{n\ln 2}{t}\right), \tag{7.4.2}$$

where the weighting factors V_n are defined by

$$V_n = (-1)^n \sum_{k=(n+1)/2}^{\min(n,N)} \frac{k^N(2k)!}{(N-k)!k!(k-1)!(n-k)!(2k-n)!}, \tag{7.4.3}$$

and the total number of terms in the series is take to be $2N$.

An advantage of the Stehfest algorithm it that it allows us to avoid evaluating $\widehat{f}(s)$ at complex values of s, as would be required by the path of integration shown in Eq. (7.1.26); instead, the method requires only that $\widehat{f}(s)$ be evaluated at real values of s.

Note, however, that if we could perform the inversion integration *analytically*, we would arrive at a single mathematical expression that would be valid for all values of t. But, if we use a *numerical* procedure, such as the Stehfest algorithm, we must perform a new calculation for *each* value of t. This is a drawback that is shared by all numerical methods.

In principle, the accuracy of the approximation should increase as the number of terms in the series, which is $2N$, increases. In practice, if too many terms are taken, accumulated round-off error begins to overshadow the additional accuracy. It has generally been found that an optimum value of $2N$ is about 18.

Problems for Chapter 7

Problem 7.1. What is the Laplace transform of the function $f(t) = e^{-\mathrm{at}}$?

Problem 7.2. Starting with the basic definition of the Laplace transform, Eq. (7.1.1), verify Eq. (7.1.16).

Problem 7.3. Using the various general properties of Laplace transforms, derive Eq. (7.1.19), $L\{t^n\} = n!/s^{n+1}$, where n is any non-negative integer.

Problem 7.4. Following the steps that were taken in Section 7.2, use Laplace transforms to solve the problem of linear flow into a hydraulic fracture with *constant pressure* in the fracture:

$$\text{PDE:} \quad \frac{1}{D}\frac{dP}{dt} = \frac{d^2P}{dz^2}, \tag{i}$$

$$\text{IC:} \quad P(z, t = 0) = P_i, \tag{ii}$$

$$\text{far-field BC:} \quad P(z \to \infty, t) = P_i, \tag{iii}$$

$$\text{fracture BC:} \quad P(z = 0, t) = P_f. \tag{iv}$$

First, find the pressure function in the Laplace domain, $\widehat{P}(z, s)$. Next, find an expression for the flow rate into the fracture, in the Laplace domain; call it $\widehat{Q}_f(s)$. Lastly, invert $\widehat{Q}_f(s)$ to find the flow rate into the fracture as a function of time, $Q_f(t)$.

Chapter 8

Naturally-Fractured Reservoirs

Many reservoirs contain a network of interconnected, naturally occurring fractures, which provide most of the reservoir-scale permeability. In fact, it has been estimated that about 40% of the world's known reserves reside in naturally fractured reservoirs. Fluid flow through such reservoirs is more complicated than through unfractured reservoirs, and cannot be accurately modelled with the equations that we have developed and solved in the previous chapters. In this chapter, we will derive the governing equations for the "dual-porosity" model that is widely used for naturally fractured reservoirs, and present the "line source" solution for flow to a vertical well in a dual-porosity reservoir.

8.1. Dual-porosity Model of Barenblatt *et al.*

The basic mathematical model used for fractured reservoirs, referred to as the *dual-porosity model*, was first developed by Barenblatt *et al.* (1960), and was used by Warren and Root (1963) to solve the problem of flow to a well in an unbounded fractured reservoir. Several refinements have been made since then, and the resulting equations have been solved for many different reservoir geometries, with and without wellbore storage, skin effects, etc. In this section, we will briefly describe this model, and present and discuss the solution for flow to a well in an unbounded naturally fractured reservoir.

In the dual-porosity model, the "macroscopic-scale" permeability of the reservoir is assumed to be provided solely by the fracture network. Fluid can only flow into the well via those fractures that are connected to the well. The regions of rock between the fractures, called "matrix blocks", contain most of the fluid, and feed this fluid into the fractures, but do not directly transmit fluid to the well.

The first step in constructing a dual-porosity model for fractured reservoirs is to start with a pressure diffusion equation for the porous medium that is *formed by the fracture network*. An obvious and straightforward modification to the derivation given in Chapter 1 shows that, in general, the diffusion equation can include a source-sink term, that provides fluid to, or takes fluid away from, the fracture system.

In the present case of a dual-porosity system, the source-sink term will represent the volumetric flux of fluid *from* the matrix blocks *into* the fractures, per unit time and per unit volume of the reservoir. Hence, in radial coordinates, the pressure diffusion equation for the "fracture system" takes the form

$$(\phi c)_f \frac{dP_f}{dt} = \frac{k_f}{\mu} \frac{1}{R} \frac{d}{dR}\left(R \frac{dP_f}{dR}\right) + q_{\mathrm{mf}}, \tag{8.1.1}$$

where the subscript f denotes the properties of the fracture network. The "fracture-matrix interaction term" q_{mf} in Eq. (8.1.1) has dimensions of $(\mathrm{m}^3/\mathrm{m}^3\mathrm{s})$, or $(1/\mathrm{s})$. The compressibility term is the total compressibility of the fluid-filled fracture system, and includes the contributions of both the formation compressibility (of the fractured rock mass) and the fluid compressibility.

The continuum-scale permeability of the fracture *network*, k_f, is related to, but is by no means equal to, the "permeability" of the individual fractures; see Section 12.8 of Jaeger *et al.* (2007). Roughly speaking, if the fractures each have aperture h, and the spacing between fractures is S, the "permeability" of any individual fracture will be $k_{if} = h^2/12$, the permeability of the fracture network will be roughly given by $k_f = h^3/12S$, and so the ratio of the permeability of the fracture network to the permeability of an individual fracture is on the order of h/S. Typical values of h are 100–1,000 μm, and typical fracture spacings are on the order of 10–100 cm, so this permeability ratio will be in the range of 10^{-4}–10^{-2}. This point is emphasised mainly to help avoid confusion when reading the literature.

Note: In a dual-porosity model, every "point" R is supposed to represent a representative elementary volume (REV) (see Section 1.3) that is large enough to encompass several fractures and matrix blocks. Having said this, such considerations are rarely invoked in practice, and dual-porosity models have been used in geothermal reservoirs, for example, where fracture spacings are as large as tens of metres!

The fracture-matrix flow term could be found by solving a flow equation within the matrix blocks, as has been done by Kazemi (1969) and others. However, most dual-porosity formulations follow Barenblatt *et al.* (1960) and assume that the flow from the matrix blocks to the fractures, at point R in the reservoir, is proportional to the difference between the pressure in the fractures and the *average* pressure in the matrix blocks, $\bar{P}_m$; this assumption is called "pseudo-steady-state fracture/matrix flow". Since the flow is expected to be proportional to the permeability of the matrix, and inversely proportional to the fluid viscosity, q_{mf} is assumed to have the form

$$q_{\mathrm{mf}} = \frac{\alpha k_m}{\mu}(\bar{P}_m - P_f), \tag{8.1.2}$$

where α is a "shape factor" that accounts for the size and shape of the matrix block.

Dimensional analysis shows that α must have dimensions of $(1/L^2)$, or $(1/\mathrm{m}^2)$ in SI units. Numerous expressions for shape factors have appeared in the literature, usually based on rough finite-difference approximations to flow within the block. For example, for a cubic block of side L, Warren and Root (1963) suggested $\alpha = 60/L^2$, Kazemi *et al.* (1976) suggested $\alpha = 12/L^2$, and Quintard and Whitaker (1996) suggested $\alpha = 49.62/L^2$.

The correct value of α for a matrix block of a given shape can be found by calculating the smallest eigenvalue of the pressure diffusion equation inside the matrix block, with constant-pressure outer boundary conditions, as shown by Zimmerman *et al.* (1993). The shape factors for a few simple, idealised geometries are as follows:

- spherical block of radius a:

$$\alpha = \frac{\pi^2}{a^2}, \tag{8.1.3}$$

- long cylindrical block of radius a:

$$\alpha = \frac{5.78}{a^2}, \tag{8.1.4}$$

- cubical block of side L:

$$\alpha = \frac{3\pi^2}{L^2}, \tag{8.1.5}$$

- rectangular block of sides $\{L_x, L_y, L_z\}$:

$$\alpha = \pi^2 \left[\frac{1}{L_x^2} + \frac{1}{L_y^2} + \frac{1}{L_z^2} \right]. \tag{8.1.6}$$

If the fracture system in the reservoir consists of three mutually orthogonal sets of parallel fractures, with spacings $\{L_x, L_y, L_z\}$, then the matrix blocks would be rectangular slabs. Hence, this latter model, although simplified, is not unrealistic.

Equations (8.1.1) and (8.1.2) give two equations for the three unknowns, $\{P_f, \bar{P}_m, q_{\mathrm{mf}}\}$. A third equation is found by performing a mass-balance on the matrix blocks. In analogy with Eq. (6.4.16), we find

$$q_{\mathrm{mf}} = -(\phi c)_m \frac{d\bar{P}_m}{dt}, \tag{8.1.7}$$

where the compressibility term represents the total compressibility of the matrix block, including both the matrix formation compressibility, and the fluid compressibility. Equations (8.1.1), (8.1.2) and (8.1.7) give a complete set of equations that can be solved to obtain the unknown pressures and flow rates.

8.2. Dual-porosity Equations in Dimensionless Form

Before presenting the solution for the pressures at the well in an unbounded dual-porosity reservoir, under conditions of constant flow rate into the well, it is convenient to first write the governing equations in dimensionless form. We first define the following dimensionless variables:

- Dimensionless time:

$$t_D = \frac{k_f t}{(\phi_f c_f + \phi_m c_m)\mu R_w^2}; \tag{8.2.1}$$

  this is the "standard" definition, but based on the permeability of the fracture network, which is k_f, and the *total* storativity, which is the sum of *fracture storativity* plus the *matrix storativity.*
- Dimensionless pressure in the fractures:

$$P_{\mathrm{Df}} = \frac{2\pi k_f H(P_i - P_f)}{\mu Q}; \tag{8.2.2}$$

  this is again the standard definition, as in Section 2.2, where Q is again the flow rate into the well.
- Dimensionless pressure in the matrix blocks:

$$P_{\mathrm{Dm}} = \frac{2\pi k_f H(P_i - \bar{P}_m)}{\mu Q}; \tag{8.2.3}$$

  this definition is based on k_f, rather than k_m, so as to be consistent with the definition of P_{Df}.
- Dimensionless radius:

$$R_D = \frac{R}{R_w}; \tag{8.2.4}$$

  which is the standard definition that we used previously for single-porosity reservoirs.

We now use these definitions, along with the chain rule, to transform Eqs. (8.1.1), (8.1.2) and (8.1.7) into dimensionless form.

We start with the left-hand side of Eq. (8.1.1):

$$\begin{aligned}(\phi c)_f \frac{dP_f}{dt} &= (\phi c)_f \frac{dP_f}{dP_{\mathrm{Df}}} \cdot \frac{dP_{\mathrm{Df}}}{dt_D} \cdot \frac{dt_D}{dt} \\ &= (\phi c)_f \left[\frac{-\mu Q}{2\pi k_f H}\right] \frac{dP_{\mathrm{Df}}}{dt_D} \left[\frac{k_f}{(\phi_f c_f + \phi_m c_m)\mu R_w^2}\right] \\ &= -\left[\frac{\phi_f c_f Q}{(\phi_f c_f + \phi_m c_m) 2\pi H R_w^2}\right] \frac{dP_{\mathrm{Df}}}{dt_D}. \end{aligned} \tag{8.2.5}$$

Applying the same procedure to the spatial derivative term in Eq. (8.1.1) gives

$$\frac{k_f}{\mu} \frac{1}{R} \frac{d}{dR}\left(R \frac{dP_f}{dR}\right) = \frac{-Q}{2\pi H R_w^2} \frac{1}{R_D} \frac{d}{dR_D}\left(R_D \frac{dP_{\mathrm{Df}}}{dR_D}\right). \tag{8.2.6}$$

Lastly, we transform the term q_{mf} in Eq. (8.1.1), using Eq. (8.1.7). In analogy with Eq. (8.2.5), this term becomes

$$q_{\mathrm{mf}} = -(\phi c)_m \frac{d\bar{P}_m}{dt} = \left[\frac{\phi_m c_m Q}{(\phi_f c_f + \phi_m c_m) 2\pi H R_w^2}\right] \frac{dP_{\mathrm{Dm}}}{dt_D}. \tag{8.2.7}$$

Inserting Eqs. (8.2.5–8.2.7) into Eq. (8.1.1) yields

$$\begin{aligned}&\frac{1}{R_D} \frac{d}{dR_D}\left(R_D \frac{dP_{\mathrm{Df}}}{dR_D}\right) \\ &\quad = \frac{\phi_f c_f}{(\phi_f c_f + \phi_m c_m)} \frac{dP_{\mathrm{Df}}}{dt_D} + \frac{\phi_m c_m}{(\phi_f c_f + \phi_m c_m)} \frac{dP_{\mathrm{Dm}}}{dt_D}. \end{aligned} \tag{8.2.8}$$

Equation (8.2.8) is the dimensionless form of Eq. (8.1.1).

We now write Eq. (8.1.2) in terms of the dimensionless variables:

$$q_{\mathrm{mf}} = \frac{\alpha k_m}{\mu}(\bar{P}_m - P_f) = \frac{-\alpha k_m Q}{2\pi k_f H}(P_{\mathrm{Dm}} - P_{\mathrm{Df}}). \tag{8.2.9}$$

Equating this result to Eq. (8.2.7) gives

$$\frac{\phi_m c_m}{(\phi_f c_f + \phi_m c_m)} \frac{dP_{\mathrm{Dm}}}{dt_D} = \frac{\alpha k_m R_w^2}{k_f}(P_{\mathrm{Df}} - P_{\mathrm{Dm}}). \tag{8.2.10}$$

Equations (8.2.8) and (8.2.10) are two coupled differential equations for the two dimensionless pressures, P_{Df} and P_{Dm}.

These two equations contain two dimensionless parameters. The first is the ratio of fracture storativity to total (fracture + matrix) storativity, denoted by ω:

$$\omega = \frac{\phi_f c_f}{\phi_f c_f + \phi_m c_m}. \tag{8.2.11}$$

The second dimensionless parameter, the transmissivity ratio λ, is essentially a ratio of matrix permeability to fracture permeability, modified by some geometrical factors:

$$\lambda = \frac{\alpha k_m R_w^2}{k_f}. \tag{8.2.12}$$

Since the fracture porosity is typically much less than the matrix porosity, and the fracture permeability is much greater than the matrix permeability, in practice it is usually the case that $\omega < 0.1$ and $\lambda < 0.001$.

In terms of the dimensionless parameters ω and λ, and the dimensionless variables defined in Eqs. (8.1.1–8.1.4), the two governing equations for radial flow in a dual-porosity reservoir are

$$\frac{1}{R_D}\frac{d}{dR_D}\left(R_D\frac{dP_{Df}}{dR_D}\right) = \omega\frac{dP_{\mathrm{Df}}}{dt_D} + (1-\omega)\frac{dP_{\mathrm{Dm}}}{dt_D}, \tag{8.2.13}$$

$$(1-\omega)\frac{dP_{\mathrm{Dm}}}{dt_D} = \lambda(P_{\mathrm{Df}} - P_{\mathrm{Dm}}). \tag{8.2.14}$$

If the reservoir were actually a "single-porosity" reservoir, then the storativity ratio ω would equal 1, and the two equations (8.2.13) and (8.2.14), would reduce to the standard pressure diffusion Eq. (6.2.8). This situation occurs in some fractured reservoirs that have very low matrix porosity and/or low matrix permeability, in which case the matrix blocks can be ignored.

In the usual form of the dual-porosity model, fluid does not flow from one matrix block to another. Only the fractures provide the macroscopic, reservoir-scale permeability. Hence, the pressure that we measure at the wellbore represents the pressure in those fractures that are nearest to the wellbore. Therefore, the drawdown at the wellbore is found from

$$\Delta P_{\rm Dw}(t_D) \equiv P_{\rm Df}(R_D = 1, t_D). \tag{8.2.15}$$

A more general model for naturally fractured reservoirs, in which fluid can also flow from one matrix block to another, is known as the "dual-permeability model". This model is sometimes used by hydrologists, but is not often used in petroleum engineering.

8.3. Line Source Problem in a Dual-porosity Reservoir

Warren and Root (1963) solved the problem of a well producing at constant flow rate Q in an unbounded dual-porosity reservoir, using Laplace transforms. They found that, if the dimensionless time satisfies the condition

$$t_D > 100\ \omega, \tag{8.3.1}$$

which usually covers the times of interest to petroleum engineers, the dimensionless pressure drawdown at the well is described by

$$\Delta P_{\rm Dw} = \frac{1}{2}\left\{\ln t_D + 0.8091 + Ei\left[\frac{-\lambda t_D}{\omega(1-\omega)}\right] - Ei\left[\frac{-\lambda t_D}{(1-\omega)}\right]\right\}. \tag{8.3.2}$$

As usual, it is instructive to analyse the behaviour of the solution in the various time regimes. In the early-time regime when $t_D > 100\ \omega$, but is not "too large", the variables inside the Ei functions in Eq. (8.3.2) will still be small enough to use Eq. (2.4.4), which says that, if $x < 0.01$,

$$-Ei(-x) \approx -0.5772 - \ln x. \tag{8.3.3}$$

Use of Eq. (8.3.3) in Eq. (8.3.2) yields

$$\begin{aligned}
\Delta P_{\rm Dw} &= \frac{1}{2}\left\{\ln t_D + 0.8091 + \ln\left[\frac{\lambda t_D}{\omega(1-\omega)}\right] - \ln\left[\frac{\lambda t_D}{(1-\omega)}\right]\right\} \\
&= \frac{1}{2}\left\{\ln t_D + 0.8091 + \ln\left[\frac{\lambda t_D}{(1-\omega)}\right] - \ln\omega - \ln\left[\frac{\lambda t_D}{(1-\omega)}\right]\right\} \\
&= \frac{1}{2}\left\{\ln(t_D/\omega) + 0.8091\right\}.
\end{aligned} \tag{8.3.4}$$

If we use Eqs. (8.2.1), (8.2.2) and (8.2.11) to convert the drawdown to *dimensional* form, we find

$$\frac{2\pi k_f H \Delta P_w}{\mu Q}$$

$$= \frac{1}{2}\left\{\ln\left[\frac{k_f t}{[(\phi c)_f + (\phi c)_m]\mu R_w^2} \cdot \frac{[(\phi c)_f + (\phi c)_m]}{(\phi c)_f}\right] + 0.8091\right\}$$

$$\text{i.e. } \Delta P_w = \frac{\mu Q}{4\pi k_f H}\left\{\ln\left[\frac{k_f t}{(\phi\mu c)_f R_w^2}\right] + 0.8091\right\}, \qquad (8.3.5)$$

which is precisely the drawdown that would occur in a single-porosity system consisting only of the fractures (i.e. without matrix blocks)! Note also that in the early-time regime, the drawdown will be a straight line on a semi-log plot.

The physical explanation of this behaviour is as follows. At early times, fluid flows to the well only through the fractures; fluid has not yet had time to flow out of the matrix blocks because of their relatively low permeability. Hence, in this regime, the matrix storativity is irrelevant.

Now let us consider "very large" times (but still assuming an unbounded reservoir). Recall from Table 2.1 that when x is large, $Ei(-x) \to 0$. So, in this regime the two Ei terms in Eq. (8.3.2) drop out, and the drawdown is

$$\Delta P_{\text{Dw}} = \frac{1}{2}\{\ln t_D + 0.8091\}, \qquad (8.3.6)$$

which in dimensional form is

$$\Delta P_w = \frac{\mu Q}{4\pi k_f H}\left\{\ln\left(\frac{k_f t}{[(\phi c)_f + (\phi c)_m]\mu R_w^2}\right) + 0.8091\right\}. \qquad (8.3.7)$$

This is the drawdown that would occur in a single-porosity reservoir whose permeability is that of the fracture system, but whose storativity is that of the fractures *plus* the matrix blocks.

The physical explanation of this behaviour is (essentially) as follows. At very large times, the matrix blocks have had sufficient time for their pressures to "equilibrate" with the pressure in the fractures. So at large times, the reservoir behaves like a homogeneous, single-porosity reservoir whose storativity is composed of the fracture storativity plus the matrix storativity. The overall permeability of the reservoir is always the permeability of the fracture system, because the matrix blocks are assumed not to be interconnected to each other.

Both the early-time solution given by Eq. (8.3.5), and the late-time solution given by Eq. (8.3.7), will give straight lines when plotted on a semi-log plot, with exactly the same slope,

$$\frac{d\Delta P_w}{d\ln t} = \frac{\mu Q}{4\pi k_f H}. \tag{8.3.8}$$

The vertical offset of these two lines, denoted by $\delta\Delta P_{\mathrm{Dw}}$, can be found by comparing Eqs. (8.3.4) and (8.3.6):

$$\delta\Delta P_{\mathrm{Dw}} = -\frac{1}{2}\ln\omega = \frac{1}{2}\ln(1/\omega). \tag{8.3.9}$$

In dimensional form, the vertical offset is

$$\delta P_w = \frac{\mu Q}{4\pi k_f H}\ln(1/\omega). \tag{8.3.10}$$

The full curve for the drawdown in a well in an infinite dual-porosity reservoir is shown in schematic form in Figure 8.1, adapted from Gringarten (1984). This figure shows that there is a transition regime between the two semi-log straight lines, in which the pressure

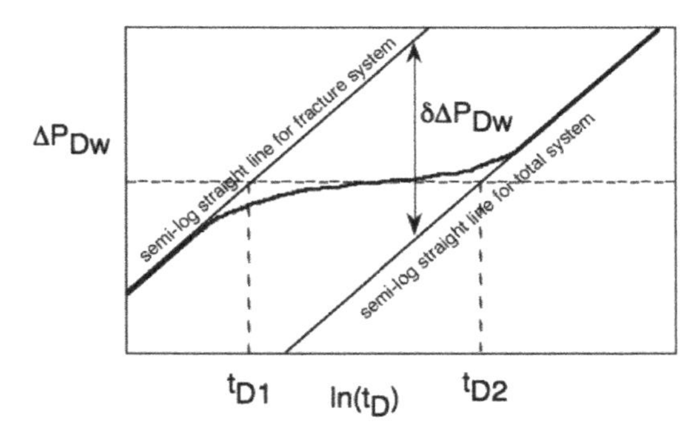

Figure 8.1. Drawdown in a well in an unbounded dual-porosity reservoir.

drawdown changes very gradually. This regime is dominated by the effect of fluid beginning to flow out of the matrix blocks and into the fractures. During this regime, the flow from the matrix blocks into the fractures nearly compensates for the flow from the fractures into the borehole, and the wellbore pressure is (nearly) constant.

Bourdet and Gringarten (1980) showed that if a horizontal line is drawn through the drawdown curve at the midpoint between the two semi-log straight lines, it will intersect the first semi-log straight line at a dimensionless time t_{D1} that is given by

$$t_{D1} = \frac{\omega}{\gamma\lambda}, \tag{8.3.11}$$

where $\gamma = 1.781$, and ω and λ are the two dual-porosity parameters defined by Eqs. (8.2.11) and (8.2.12). This horizontal line intersects the late-time semi-log straight line at a dimensionless time t_{D2} given by

$$t_{D2} = \frac{1}{\gamma\lambda}. \tag{8.3.12}$$

Equations (8.3.11) and (8.3.12) show that the horizontal offset of the two asymptotic semi-log straight lines is equal to $\ln(1/\omega)$. These two equations provide a simple means to estimate the value of the storativity ratio ω, and the transmissivity ratio λ, from a well test.

In dimensional form, these intercept times are

$$t_1 = \frac{(\phi\mu c)_f}{\alpha\gamma k_m}, \tag{8.3.13}$$

$$t_2 = \frac{\mu[(\phi c)_f + (\phi c)_m]}{\alpha\gamma k_m}. \tag{8.3.14}$$

The first intercept time, t_1, is essentially a measure of the time required for the volume of fluid that has been depleted from the matrix blocks near the wellbore to become of the same order of magnitude as the volume depleted from the fractures. The second intercept time, t_2, is a measure of the time required for the pressure in the matrix blocks nearest the well to come into equilibrium with the pressure in the surrounding fractures.

Problems for Chapter 8

Problem 8.1. Without looking at the paper by Warren and Root, describe and sketch the way that the drawdown curve in Figure 8.1 would change if (a) the storativity ratio ω increased (or decreased) by a factor of 10, or (b) the transmissivity ratio λ increased (or decreased) by a factor of 10.

Problem 8.2. By examining Eq. (8.3.2), and making use of either Eq. (2.1.22) or Table 2.1, derive an expression for the dimensionless time that must elapse in order for the approximation (8.3.6) to be accurate to within about 1%. Your answer should be expressed in terms of the parameter λ.

Chapter 9

Flow of Gases in Porous Media

Most of the equations derived and analysed in previous chapters are applicable only when the flowing fluid is a liquid, in which case the compressibility and viscosity can be assumed to be constant, and independent of pressure. For gas reservoirs, the assumption that these parameters are independent of pressure is not very accurate, and incorporation of their pressure dependence into the pressure diffusion equation renders it nonlinear. In this chapter, we will show how the resulting nonlinear diffusion equation for gas flow can be (approximately) linearised, by expressing it in terms of a new variable known as the pseudo-pressure.

Another important assumption made in all previous chapters was the validity of Darcy's law, which states that the flow rate is linearly proportional to the pressure gradient. However, this "law" becomes inaccurate at high flow rates, where it is usually replaced by a nonlinear relation between flow rate and pressure gradient known as the Forchheimer equation. Although this deviation from Darcy's law occurs for both liquids and gases, flow rates in gas reservoirs are typically higher, due to the lower viscosity of gases, and so it is much more likely that flow in a gas reservoir will take place in this non-Darcy regime.

Deviations from Darcy's law can also occur for gas flow if the pressure is sufficiently low, relative to some characteristic pressure that is inversely proportional to the mean pore size. In this situation, the "effective gas permeability" differs from the true permeability, due to slippage of the gas molecules along the pore walls. This so-called "Klinkenberg effect" must be taken into account when analysing laboratory permeability measurements performing using gas at low pressures, and also for gas flow (at any pressure) in ultra-low permeability formations such as shales.

9.1. Diffusion Equation for Gas Flow in Porous Media

The main difference between analysing flow in gas reservoirs as opposed to liquid reservoirs is that, for gases, the governing partial differential equation becomes unavoidably *nonlinear.* The main reason for this nonlinearity is that, whereas the compressibility of a liquid can be assumed to be constant, and relatively small, in the sense that was quantified in Section 1.6, the compressibility of a gas varies strongly with pressure. Furthermore, the viscosity of a gas usually also varies with pressure. The result is that the governing equation for gas flow cannot usually be well approximated by a linear diffusion equation with constant coefficients, as was the case for liquids.

To derive the governing equation for gas flow, we return to the analysis given in Section 1.7, and note that, up to Eq. (1.7.5), no assumptions were made about the magnitude of the compressibility. Hence, we can begin our analysis of gas flow with Eq. (1.7.5):

$$-\frac{d(\rho q R)}{dR} = R\frac{d(\rho\phi)}{dt}. \tag{9.1.1}$$

We now use Darcy's law, in the form given by Eq. (1.4.1), for the flux term q on the left-hand side, i.e.

$$q = -\frac{k}{\mu}\frac{dP}{dR}, \tag{9.1.2}$$

in which case Eq. (9.1.1) takes the form

$$\frac{1}{R}\frac{d}{dR}\left(\frac{\rho k}{\mu}R\frac{dP}{dR}\right) = \frac{d(\rho\phi)}{dt}. \tag{9.1.3}$$

We next note that the process of expressing the right-hand side of Eq. (9.1.3) in terms of the pressure, which was carried out in

Section 1.6, did *not* require any assumptions about the magnitude of the compressibility terms, etc. Hence, we can use Eq. (1.6.1) to rewrite Eq. (9.1.3) as

$$\frac{1}{R}\frac{d}{dR}\left(\frac{\rho k}{\mu}R\frac{dP}{dR}\right) = \rho\phi(c_f + c_\phi)\frac{dP}{dt}. \tag{9.1.4}$$

Equation (9.1.4) is completely general, in that it contains no assumption that the compressibility of the pore fluid is small. It also allows for the possibility that the density, permeability and viscosity vary with pressure. As such, it is too general to be useful, except as the basis of a numerical solution. However, there are many cases of practical interest in which it can be either linearised, or "almost" linearised, after which the standard solutions presented in the previous chapters can be used.

9.2. Ideal Gas, Constant Reservoir Properties

One situation in which the nonlinear diffusion equation for gas flow can be effectively linearised is if the gas can be assumed to obey the ideal gas law:

$$P = \rho \mathbf{R} T, \tag{9.2.1}$$

where $\mathbf{R}$ is the gas constant, and T is the absolute temperature. If we use Eq. (9.2.1) to express the density in terms of the pressure, Eq. (9.1.4) takes the form

$$\frac{1}{R}\frac{d}{dR}\left(\frac{kP}{\mu \mathbf{R} T}R\frac{dP}{dR}\right) = \frac{\phi(c_f + c_\phi)}{\mathbf{R}T}P\frac{dP}{dt}. \tag{9.2.2}$$

The simplest model of gas flow can be derived from Eq. (9.2.2) by making the following assumptions:

(a) The temperature is constant.
(b) The viscosity of the gas is independent of pressure. This is rigorously true for an ideal gas, for which the viscosity depends only on temperature. This assumption allows us to take μ outside of the derivative on the left-hand side of Eq. (9.2.2).

(c) The compressibility of the gas is much larger than that of the formation, which is often the case for gas reservoirs. In fact, the compressibility of an ideal gas is $1/P$, since

$$c_f = \frac{1}{\rho}\left(\frac{d\rho}{dP}\right)_T = \frac{\mathbf{R}T}{P}\frac{d}{dP}\left(\frac{P}{\mathbf{R}T}\right)_T = \frac{\mathbf{R}T}{P}\frac{1}{\mathbf{R}T} = \frac{1}{P}. \quad (9.2.3)$$

As the reservoir pressure is usually less than 10,000 psi, the compressibility is at least 10^{-4}/psi. The formation compressibility, on the other hand, is usually on the order of 10^{-5}/psi (Matthews and Russell, 1967; Zimmerman, 1991). Hence, we can often neglect c_ϕ relative to c_f.

(d) Assume that the permeability of the formation is independent of the pore pressure. This is reasonably accurate for many reservoirs, but is sometimes *not* a very good assumption in fractured reservoirs or in tight gas sands, for example. This assumption allows us to take k outside of the derivative on the left-hand side of Eq. (9.2.2).

Under these four assumptions, Eq. (9.2.2) can be written as

$$\frac{1}{R}\frac{d}{dR}\left(RP\frac{dP}{dR}\right) = \frac{\phi\mu}{kP}P\frac{dP}{dt}. \quad (9.2.4)$$

We now note that

$$P\frac{dP}{dt} = \frac{1}{2}\frac{d(P^2)}{dt}, \quad P\frac{dP}{dR} = \frac{1}{2}\frac{d(P^2)}{dR}, \quad (9.2.5)$$

which allows us to write Eq. (9.2.4) as

$$\frac{1}{R}\frac{d}{dR}\left[R\frac{d(P^2)}{dR}\right] = \frac{\phi\mu}{kP}\frac{d(P^2)}{dt}. \quad (9.2.6)$$

Equation (9.2.6) is the standard diffusion equation in radial coordinates, *except* that

(a) The dependent variable is P^2 rather than P;
(b) The compressibility term, $c_t \approx c_f = 1/P$, varies with pressure.

Because of point (b), Eq. (9.2.6) is still nonlinear, and is not quite equivalent to the diffusion equation used for liquid flow. However, we can linearise Eq. (9.2.6) by evaluating the term P in the denominator of the RHS at either the initial pressure, P_i, or at the mean wellbore pressure during the test, i.e. at

$$P_m = \frac{P_i + P_w(\text{end of test})}{2}. \tag{9.2.7}$$

If this is done, then Eq. (9.2.6) becomes a linear diffusion equation, with P^2 rather than P as the dependent variable.

9.3. Real Gas, Variable Reservoir Properties

The ideal gas law is a good approximation to gas behaviour only at low pressures. It is also generally more accurate for monatomic or diatomic gases than for larger molecules such as gaseous hydrocarbons. For gas reservoirs, which are usually at relatively high pressures, it is more accurate to replace the ideal gas law, Eq. (9.2.1), with the "real gas" equation of state,

$$P = \rho z \mathbf{R} T, \quad \text{or} \quad \rho = \frac{P}{z\mathbf{R}T}, \tag{9.3.1}$$

where z is the (dimensionless) "gas deviation factor". Using this equation of state, Eq. (9.1.3) takes the form

$$\frac{1}{R}\frac{d}{dR}\left(R\frac{k}{\mu z \mathbf{R} T} P \frac{dP}{dR}\right) = \phi \frac{d}{dt}\left(\frac{P}{z\mathbf{R}T}\right). \tag{9.3.2}$$

Under isothermal conditions, this becomes

$$\frac{1}{R}\frac{d}{dR}\left(R\frac{k}{\mu z} P \frac{dP}{dR}\right) = \phi \frac{d}{dt}\left(\frac{P}{z}\right). \tag{9.3.3}$$

Equation (9.3.3) is highly nonlinear, but we can "partially linearise" it by defining a new variable, the "real gas pseudopressure", as follows; see Chapter 7 of *Principles of Petroleum Reservoir Engineering*, by Chierici (1994):

$$m(P) = 2\int_0^P \frac{k(P)P}{\mu(P)z(P)}dP. \tag{9.3.4}$$

The real gas pseudopressure $m(P)$ is just a generalisation of the P^2 parameter that was used for ideal gases. This can be seen by noting that if we assume that the permeability and viscosity are both independent of pressure, and recall that $z = 1$ for an ideal gas, then

$$m(P) \to \frac{2k}{\mu}\int_0^P PdP = \frac{k}{\mu}P^2. \tag{9.3.5}$$

Hence, for an ideal gas in a stress-*in*sensitive reservoir, the pseudopressure $m(P)$ is, aside from a multiplicative constant, equal to P^2.

Returning to the real gas case, we differentiate Eq. (9.3.4) with respect to P, to find

$$\frac{dm(P)}{dP} = \frac{2k(P)P}{\mu(P)z(P)}, \tag{9.3.6}$$

which implies that

$$\frac{dm}{dR} = \frac{dm}{dP}\frac{dP}{dR} = \frac{2k(P)P}{\mu(P)z(P)}\frac{dP}{dR}. \tag{9.3.7}$$

Substituting Eq. (9.3.7) into the left-hand side of Eq. (9.3.3) leads to

$$\frac{1}{2R}\frac{d}{dR}\left(R\frac{dm}{dR}\right) = \phi\frac{d}{dt}\left(\frac{P}{z}\right). \tag{9.3.8}$$

The left-hand side of Eq. (9.3.8) is essentially in the standard form for the diffusion equation. On the right-hand side, we recall that $P/z = \rho\mathbf{R}T$, so that under isothermal conditions,

$$\frac{d}{dt}\left(\frac{P}{z}\right) = \frac{d(\rho\mathbf{R}T)}{dt} = \mathbf{R}T\frac{d\rho}{dt} = \mathbf{R}T\frac{d\rho}{dP}\frac{dP}{dt}. \tag{9.3.9}$$

But from the definition of compressibility,

$$c_g = \frac{1}{\rho}\left(\frac{d\rho}{dP}\right)_T, \quad \text{so} \quad \left(\frac{d\rho}{dP}\right)_T = \rho c_g, \tag{9.3.10}$$

in which case Eq. (9.3.9) can be written as

$$\frac{d}{dt}\left(\frac{P}{z}\right) = \mathbf{R}T\rho(P)c_g(P)\frac{dP}{dt}. \tag{9.3.11}$$

Next we recall from Eq. (9.3.1) that $\rho = P/z\mathbf{R}T$, and rewrite Eq. (9.3.11) as

$$\frac{d}{dt}\left(\frac{P}{z}\right) = \frac{c_g(P)P}{z(P)}\frac{dP}{dt}. \tag{9.3.12}$$

Combining Eq. (9.3.12) with Eq. (9.3.8) yields

$$\frac{1}{2R}\frac{d}{dR}\left(R\frac{dm}{dR}\right) = \frac{\phi c_g(P)P}{z(P)}\frac{dP}{dt}. \tag{9.3.13}$$

But by analogy with Eq. (9.3.7), we have

$$\frac{dm}{dt} = \frac{2k(P)P}{\mu(P)z(P)}\frac{dP}{dt}, \tag{9.3.14}$$

so that

$$\frac{dP}{dt} = \frac{\mu(P)z(P)}{2k(P)P}\frac{dm}{dt}. \tag{9.3.15}$$

Combining Eq. (9.3.15) with Eq. (9.3.13) yields

$$\frac{1}{R}\frac{d}{dR}\left(R\frac{dm}{dR}\right) = \frac{\phi\mu(P)c_g(P)}{k(P)}\frac{dm}{dt}. \tag{9.3.16}$$

Equation (9.3.16) is in the form of a standard diffusion equation in radial coordinates, except that the diffusivity term on the right-hand side is pressure-dependent.

As a final step in linearising the equation for gas flow to a well, we make the approximation that the diffusivity terms on the right can be replaced by some representative constant value, such as the value at the mean wellbore pressure encountered during a well test, i.e.

$$\frac{1}{R}\frac{d}{dR}\left(R\frac{dm}{dR}\right) = \frac{\phi\mu(P_m)c_g(P_m)}{k(P_m)}\frac{dm}{dt}, \tag{9.3.17}$$

where P_m is defined by Eq. (9.2.7).

Equation (9.3.17) is the (approximate) governing equation for the flow of real gases to a well. If the pressure-dependent terms on the right-hand side are evaluated at P_m, we can then use all of the standard solutions that have been developed for liquid flow, provided that we use m instead of P as the dependent variable.

The transformation between P and m is made by evaluating the integral that is given in Eq. (9.3.4). If the reservoir is stress-sensitive, we would need to know $k(P)$ in order to do this. Usually, we do not have this information. If we ignore the pressure-dependence of permeability, then the relationship between P and m becomes

$$m(P) = 2k \int_0^P \frac{P}{\mu(P)z(P)} dP. \tag{9.3.18}$$

Since $z(P)$ and $\mu(P)$ can be measured in the laboratory on samples of the reservoir gas, this integral can be evaluated *a priori*. The result is usually a relationship between P and m in tabular form, which is used to transform the measured pressures into values of m that are used in conjunction with our solutions to Eq. (9.3.17).

9.4. Non-Darcy Flow Effects

All of the previous analyses have been based on Darcy's law, which states that the flow rate is proportional to the pressure gradient. Darcy's law is an empirical law that is known to hold only at low flow rates, which can be defined, roughly, as flows for which the Reynolds number is less than one. The Reynolds number is a dimensionless measure of the relative strengths of inertial forces relative to viscous forces. Using the definition of Reynolds number, this condition can be written as

$$Re = \frac{\rho v d}{\mu} < 1, \tag{9.4.1}$$

where ρ is the density of the fluid, μ is the viscosity, d is a mean pore diameter and v is the mean (microscopic) velocity.

It would be useful to be able to express this criterion in terms of macroscopically measurable quantities. If we use any of the common correlations between pore diameter and permeability, such as the Kozeny–Carman equation,

$$k = \frac{\phi d^2}{96} \approx \frac{\phi d^2}{100}, \tag{9.4.2}$$

we can express the pore diameter as

$$d = 10\sqrt{k/\phi}. \tag{9.4.3}$$

The criterion for Darcy's law to be valid can therefore be written as

$$v < \frac{\mu\sqrt{\phi}}{10\rho\sqrt{k}}. \tag{9.4.4}$$

Next, we note that the macroscopic flow rate q must be equal to the microscopic flow rate v, multiplied by the porosity, i.e. $q = v\phi$, so that

$$v = \frac{q}{\phi}. \tag{9.4.5}$$

Hence, the criterion (9.4.4) for the validity of Darcy's law can be written as

$$q < \frac{\mu\phi^{3/2}}{10\rho\sqrt{k}}. \tag{9.4.6}$$

The total flow rate towards the well is related to the flux by

$$Q = qA = 2\pi RHq \tag{9.4.7}$$

and so condition (9.4.6) can be written as

$$Q < \frac{2\pi RH\mu\phi^{3/2}}{10\rho\sqrt{k}} \approx \frac{RH\mu\phi^{3/2}}{\rho\sqrt{k}}, \tag{9.4.8}$$

which is expressed solely in terms of our usual reservoir parameters.

Criterion (9.4.8) becomes more stringent near the wellbore, where R is smallest. This is because as a fixed volumetric flow rate gets channelled through a smaller area, the velocity must increase. Consequently, Darcy's law is more likely to be become inapplicable near the wellbore than farther out in the reservoir.

If we use "typical" reservoir values in Eq. (9.4.8), we would find that this criterion is usually violated in the vicinity of the wellbore for gas flow, but not for liquid flow, unless the flow is occurring through fractures rather than pores. If the flowing is occurring through fractures, then the actual area available for flow is smaller, and the velocity of the fluid must be greater. In this case, it is more likely that the flow will be "non-Darcy".

If criterion (9.4.8) is violated, Darcy's law must be replaced with a nonlinear law, such as Forchheimer's equation:

$$\frac{dP}{dR} = \frac{\mu q}{k} + \beta \rho q^2, \tag{9.4.9}$$

in which the term $\beta\rho q^2$ represents an additional pressure gradient, due to inertial effects, and β is the Forchheimer coefficient. Note that when writing Eq. (9.4.9), we assume that q is positive if the fluid is flowing *towards* the well. The Forchheimer equation can be justified heuristically by noting that ρq^2 is related to the kinetic energy per unit volume of fluid, and the kinetic energy is *not* negligible at higher flow rates (recall Eq. 1.1.2).

Equation (9.4.9) implies that, for a given flow rate, the pressure drop will be *larger* than that predicted by Darcy's law. Hence, non-Darcy effects contribute an "additional" pressure drop. Conversely, since the "non-Darcy" pressure drop is greater than the Darcy pressure drop for a given flow rate, then, if non-Darcy effects are important, the flow rate due to a given pressure drop will be *less* than that predicted by Darcy's law.

Dimensional analysis of Eq. (9.4.9) shows that the factor β has dimensions of L^{-1}. As k has dimensions of L^2 (recall Eq. 9.4.2), it is roughly the case that $\beta \approx 1/\sqrt{k}$. If this is true, then the magnitude of the nonlinear term in Eq. (9.4.9) relative to that of the linear term is

$$\text{ratio} = \frac{\beta\rho q^2}{\mu q/k} = \frac{\beta\rho q k}{\mu} \approx \frac{\rho q\sqrt{k}}{\mu}. \tag{9.4.10}$$

The term on the right of Eq. (9.4.10) is essentially a Reynolds number based on a length scale of $\sqrt{k}$. Hence, if the Reynolds number is much less than one, the nonlinear terms in Eq. (9.4.9) are negligible, and we recover Darcy's law.

As non-Darcy flow is confined to a region near the wellbore, its effect during a well test is somewhat similar to a "skin" effect. In fact, the main consequence of non-Darcy flow near the wellbore is that the skin factor gets "replaced" by an apparent skin factor, s', which is given by

$$s' = s + DQ, \tag{9.4.11}$$

where D is the "non-Darcy skin coefficient". This coefficient is related to the coefficient β that appears in Forchheimer's equation; see p. 241 of Chierici (1994) for details.

9.5. Klinkenberg Effect

Liquids, as well as gases at typical reservoir pressures, behave like continua, in the sense that we can ignore the motions of individual molecules, and instead work with mean velocities that are averaged over a (very) large number of molecules. This leads to continuum-type theories such as Darcy's law for flow through porous media, which actually arise from the Navier–Stokes equations for fluid flow at the pore scale.

A fundamental aspect of using the Navier–Stokes equations is that the fluid velocity is assumed to be zero at the boundary with a solid, such as at the pore walls. This obviously must be true for the normal component of the velocity, but it is also true for the tangential component. This "no-slip" boundary condition is the main reason why the Navier–Stokes equations average out ("upscale") to yield Darcy's law.

However, this microscale boundary condition will *not* hold for a gas at very low densities, which is to say at very low pressures. The reasons are complicated, but they boil down to the fact that in order for the gas to behave like a continuum, a given gas molecule must collide much more frequently with other gas molecules than with the pore walls. At low densities, however, each gas molecule will collide with a pore wall much more frequently than it collides with another gas molecule.

To quantify whether or not this will be the case, we must consider the concept of the "mean free path", which is essentially the mean distance travelled by a molecule between subsequent collisions with other molecules. According to the kinetic theory of gases, the mean free path λ is essentially given by (see *Molecular Theory of Gases and Liquids*, Hirschfelder *et al.* (1954)):

$$\lambda = \frac{1}{\sqrt{2}n\pi\sigma^2} = \frac{k_B T}{\sqrt{2}\pi\sigma^2 P}, \tag{9.5.1}$$

where n is the molecular density in molecules/volume, k_B is the Boltzmann constant (i.e. the gas constant per molecule), T is the absolute temperature, and σ is an effective molecular diameter.

If the pore size is smaller than the mean free path, collisions with the pore walls will be much more frequent than collisions with other molecules, and the gas will essentially flow through the pores as individual molecules, rather than as a fluid continuum. This type of flow is known as "Knudsen flow", or "slip flow".

Klinkenberg (1941) assumed that gas flow through a porous medium could be modelled as Knudsen flow through a capillary tube, and showed that the "apparent" permeability measured during gas flow will be related to the "true" absolute permeability k by

$$k_{\text{gas}} = k\left[1 + \frac{8c\lambda}{d}\right], \tag{9.5.2}$$

where λ is the mean free path, d is the pore diameter, and $c \approx 1$ is a dimensionless coefficient. If we combine Eqs. (9.5.1) and (9.5.2), we find

$$k_{\text{gas}} = k\left[1 + \frac{8c}{\sqrt{2}\pi}\frac{k_B T}{d\sigma^2}\frac{1}{P}\right]. \tag{9.5.3}$$

If we now use Eq. (9.4.3) to relate the pore diameter to the permeability, we find

$$k_{\text{gas}} = k\left[1 + \frac{4c\sqrt{\phi}}{5\sqrt{2}\pi}\frac{k_B T}{\sqrt{k}\sigma^2 P}\right]. \tag{9.5.4}$$

The second term in the brackets in Eq. (9.5.4) is the relative discrepancy due to the Klinkenberg effect. Noting that the temperature, the molecular diameter, and the term $\sqrt{\phi}$, will not vary by very much, for all cases of practical interest, we see that the main parameters that control the strength of the Klinkenberg effect are the pressure and the permeability. Equation (9.5.4) shows that the Klinkenberg effect is enhanced if either (i) the pressure is low, or (ii) the permeability is low.

For a gas flowing through a given rock at a fixed temperature, we can evaluate all the terms inside the bracket except P, and call the result P^*. Equation (9.5.4) can then be written as

$$k_{\text{gas}} = k\left[1 + \frac{P^*}{P}\right], \tag{9.5.5}$$

where P^*, which is usually written as b, is a characteristic pressure that has the following significance: the Klinkenberg effect will have a noticeable effect on the measured permeability if (roughly) $P < 10P^*$. Alternatively, we can say that the Klinkenberg effect will be negligible if $P > 10P^*$.

To quantify the magnitude of the Klinkenberg effect, consider a rock having a porosity of $\phi = 0.10$, with a gas flowing through it at 300°K. Boltzmann's constant is $1.38 \times 10^{-23}\,\text{J/°K}$, and typical molecular diameters are on the order of 4Å, so we find from Eq. (9.5.4) that

$$\text{for } k = 10\,\text{mD:}\quad P^* \approx 15\,\text{kPa} \approx 2\,\text{psi},$$

$$\text{for } k = 1000\,\text{mD:}\quad P^* \approx 1.5\,\text{kPa} \approx 0.2\,\text{psi}.$$

Reservoir pressures will be much larger than P^* for conventional reservoir rocks, and so the Klinkenberg effect is usually negligible for flow in the reservoir. Exceptions will occur in shale gas reservoirs, where k is *very* small, and P^* will be large. In shale gas reservoirs, Klinkenberg effects may be important.

Even for conventional reservoir rocks that do not have ultra-low permeabilities, Klinkenberg effects must be taken into account, for the following reason. In the laboratory, permeability tests are often conducted using gas at low pressures because such measurements are quicker, safer and cheaper than measurements made at simulated reservoir pressures. For the lab data, it will not usually be the case that the pressures are much larger than P^*. Consequently, lab-measured permeabilities conducted using a gas such as nitrogen must always be corrected to eliminate the Klinkenberg effect, and arrive at the "true" permeability. In this case, it is the actual permeability k that we want to know, *not* the apparent gas permeability, k_{gas}, which in this case is an experimental artefact!

This "Klinkenberg correction" to laboratory data is achieved by measuring k_{gas} over a range of pressures, and then plotting the results as a function of $1/P$. According to Eq. (9.5.5), the data should fall on a straight line with a positive slope. If we then fit a straight line through the k versus $1/P$ data, and extrapolate this line back to $1/P = 0$, we can find the absolute permeability, k. This procedure is illustrated in Figure 9.1, adapted from p. 78 of Chierici (1994). Note that $1/P = 0$ corresponds to very large pressures, at which the gas does behave like a continuum.

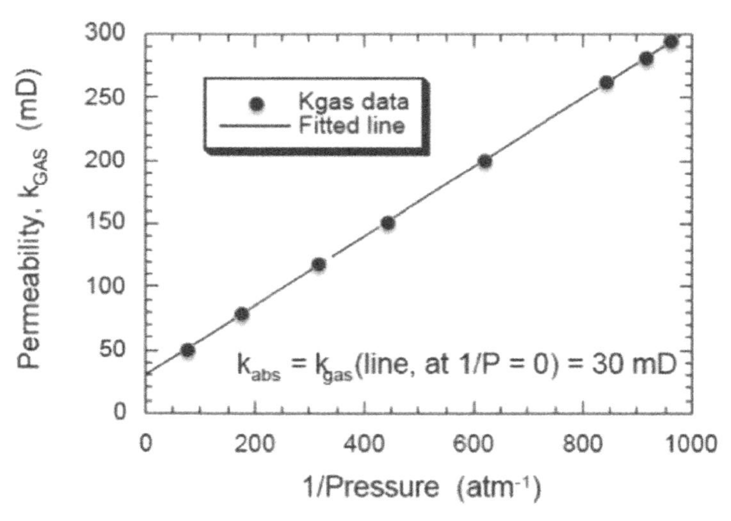

Figure 9.1. Apparent "gas permeability" at low pressures, extrapolated to $1/P = 0$ to find the true permeability.

Note that the parameter P^* contains k within it as part of its definition, and we do not know k before we measure it! However, the extrapolation procedure does *not* require knowledge of k. The data can be plotted against $1/P$ in any units, dimensionless or not; extrapolation to $1/P = 0$ will give the "actual" permeability.

Problem for Chapter 9

Problem 9.1. This problem involves a simplified model for production of gas from a hydraulically fractured shale gas reservoir.

Combining Eqs. (8.1.2) and (8.1.7) leads to the following ODE that governs the mean pressure in a matrix block in a fractured reservoir:

$$\frac{d\bar{P}_m}{dt} = \frac{-\alpha k_m}{\phi_m \mu c_m}(\bar{P}_m - P_f). \tag{P.1}$$

As mentioned in Section 9.2, in a gas reservoir the fluid compressibility term will usually greatly exceed the formation compressibility term, and so the total compressibility of the gas-filled matrix block can be approximated by $c_m \doteq c_{\text{gas}}$. Furthermore, if we approximate the behaviour of the gas as an ideal gas, then according to Eq. (9.2.3), we can say that $c_m = c_{\text{gas}} = 1/P_m$.

(a) Imagine that a matrix block in a gas reservoir is initially at some pressure P_i, after which the pressure in the surrounding fractures is lowered to some value $P_f < P_i$, and then held at that pressure. Approximate the term $c_m = 1/P_m$ by its *initial* value $1/P_i$, and then integrate Eq. (P.1) to find the mean pressure in the matrix block, as a function of time. Then use Eq. (8.1.2) to find an expression for the rate of gas production from the matrix block as a function of time.
(b) Noting that e^{-x} will be < 0.01 when $x > 5$ (roughly), derive an expression for the time at which the production rate of gas from the matrix block has diminished to 1% of its initial value.
(c) When a shale gas reservoir is hydraulically fractured, the interaction between the hydraulically induced fractures and the pre-existing microfractures and bedding planes will create a

network of fractures around the wellbore. Imagine that this fracture network breaks up the region around the wellbore into a collection of blocks that can be approximated by cubes of size L. Assuming plausible values such as $L = 1\,\text{m}$, $\phi_m = 0.1$, $\mu = 1 \times 10^{-5}\,\text{Pa s}$, $k_m = 10^{-21}\,\text{m}^2$, and $P_i = 20\,\text{MPa}$, how long will it take for the production rate to drop to 1% of its initial value?

Appendix

Solutions to Problems

Problem 1.1. A well located in a 100 ft. thick reservoir having a permeability of 100 mD produces 100 bbl/day of oil from a 10 in. diameter wellbore. The viscosity of the oil is 0.4 cP. The pressure at a distance of 1000 ft. from the wellbore is 3000 psi. What is the pressure at the wellbore? Conversion factors are as follows:

$$1 \text{ barrel} = 0.1589 \text{ m}^3,$$

$$1 \text{ Poise} = 0.1 \text{ N s/m}^2,$$

$$1 \text{ foot} = 0.3048 \text{ m},$$

$$1 \text{ psi} = 6895 \text{ N/m}^2 = 6895 \text{ Pa}.$$

Solution: The wellbore pressure is given by Eq. (1.4.5), reprinted below for convenience:

$$P_w = P_o + \frac{\mu Q}{2\pi k H} \ln\left(\frac{R_w}{R_o}\right). \tag{A.1}$$

First, convert all given data into SI units:

$$R_w = (5 \text{ in}) \left(\frac{1 \text{ ft}}{12 \text{ in}}\right) \left(\frac{0.3048 \text{ m}}{1 \text{ ft}}\right) = 0.127 \text{ m},$$

$$R_o = (1000 \text{ ft}) \left(\frac{0.3048 \text{ m}}{1 \text{ ft}}\right) = 304.8 \text{ m},$$

$$H = (100 \text{ ft}) \left(\frac{0.3048 \text{ m}}{1 \text{ ft}}\right) = 30.48 \text{ m},$$

$$\mu = (0.4\text{ cP})\left(\frac{1\text{ P}}{100\text{ cP}}\right)\left(\frac{0.1\text{ Pa s}}{1\text{ P}}\right) = 4\times10^{-4}\text{ Pa s},$$

$$k = (100\text{ mD})\left(\frac{1\text{ D}}{1000\text{ mD}}\right)\left(\frac{0.987\times10^{-12}\text{ m}^2}{1\text{ D}}\right)$$

$$= 9.87\times10^{-14}\text{ m}^2,$$

$$Q = \left(100\frac{\text{bbl}}{\text{day}}\right)\left(\frac{0.1589\text{ m}^3}{\text{bbl}}\right)\left(\frac{1\text{ day}}{24\text{ hr}}\right)\left(\frac{1\text{ hr}}{3600\text{ sec}}\right)$$

$$= 1.84\times10^{-4}\text{ m}^3/\text{s}.$$

Plugging these values into Eq. (A.1), and then converting the answer back into psi, gives

$$P_w - P_o = \frac{(4\times10^{-4}\text{ Pa s})(1.84\times10^{-4}\text{ m}^3/\text{s})\ln\left(\frac{0.127}{304.8}\right)}{2\pi(9.87\times10^{-14}\text{ m}^2)(30.48\text{ m})},$$

$$= -30310\text{ Pa}\left(\frac{1\text{ psi}}{6895\text{ Pa}}\right) = -4.4\text{ psi}.$$

$$\Rightarrow \quad P_w = P_o - 4.4\text{ psi} = 3000 - 4.4\text{ psi} = 2995.6\text{ psi}.$$

Problem 1.2. Carry out a derivation of the diffusion equation for *spherically-symmetric* flow, in analogy to the derivation given in Section 1.7 for radial flow. (This equation can be used to model flow to a well in situations when only a small length of the well has been perforated, in which case the large-scale flow field will, at early times, be roughly spherical). The result of your derivation should be an equation similar to Eq. (1.7.8), but with a slightly different term on the right-hand side.

Solution: The derivation is essentially the same as for the cylindrical geometry; the only difference being that we use a spherical shell of radius R and thickness ΔR, rather than a cylindrical shell.

So, the cross-sectional area normal to the flow will be $4\pi R^2$ rather than $2\pi RH$, and the volume of the shell will be $4\pi R^2\Delta R$ rather than $2\pi RH\Delta R$. For clarity, the derivation will be presented step-by-step, starting with Eq. (1.5.4), which is generic

$$[A(x)\rho(x)q(x) - A(x+\Delta x)\rho(x+\Delta x)q(x+\Delta x)]\Delta t$$
$$= m(t+\Delta t) - m(t). \tag{A.2}$$

Replace x with R, and note that $A(R) = 4\pi R^2$

$$[4\pi R^2\rho(R)q(R) - 4\pi(R+\Delta R)^2\rho(R+\Delta R)q(R+\Delta R)]\Delta t$$
$$= m(t+\Delta t) - m(t). \tag{A.3}$$

As before, divide by Δt, and let $\Delta t \to 0$, to find

$$4\pi[R^2\rho(R)q(R) - (R+\Delta R)^2\rho(R+\Delta R)q(R+\Delta R)] = \frac{dm}{dt}. \tag{A.4}$$

On the right side:

$$m = \rho\phi V = \rho\phi 4\pi R^2\Delta R, \tag{A.5}$$

$$\Rightarrow \quad \frac{dm}{dt} = \frac{d(\rho\phi 4\pi R^2\Delta R)}{dt} = 4\pi R^2\frac{d(\rho\phi)}{dt}\Delta R. \tag{A.6}$$

Equate Eqs. (A.4) and (A.5), divide by ΔR, and let $\Delta R \to 0$. The 4π terms cancel out, leaving

$$-\frac{d(\rho q R^2)}{dR} = R^2\frac{d(\rho\phi)}{dt}. \tag{A.7}$$

Equation (A.7) is the spherical-flow version of the conservation of mass equation.

Now use Darcy's law in the form of Eq. (1.1.5) for q on the left side of Eq. (A.7), use Eq. (1.6.1) on the right side, and then divide through by k/μ, to get

$$\frac{d}{dR}\left(\rho R^2\frac{dP}{dR}\right) = \frac{\rho\mu\phi(c_f + c_\phi)R^2}{k}\frac{dP}{dt}. \tag{A.8}$$

Expand out the left side using the product rule for derivatives, treating ρ and $R^2(dP/dR)$ as the two terms:

$$\begin{aligned}\frac{d}{dR}\left(\rho R^2\frac{dP}{dR}\right) &= \frac{d\rho}{dR}\left(R^2\frac{dP}{dR}\right)+\rho\frac{d}{dR}\left(R^2\frac{dP}{dR}\right)\\ &= \left(\frac{d\rho}{dP}\frac{dP}{dR}\right)\left(R^2\frac{dP}{dR}\right)+\rho\frac{d}{dR}\left(R^2\frac{dP}{dR}\right)\\ &= \rho c_f R^2\left(\frac{dP}{dR}\right)^2+\rho\frac{d}{dR}\left(R^2\frac{dP}{dR}\right).\end{aligned} \tag{A.9}$$

For liquids, the first term on the right of Eq. (A.9) is usually negligible compared to the second, so Eq. (A.8) becomes

$$\rho\frac{d}{dR}\left(R^2\frac{dP}{dR}\right)=\frac{\rho\mu\phi(c_f+c_\phi)R^2}{k}\frac{dP}{dt}. \tag{A.10}$$

Cancel out ρ from both sides, and rename $(c_f + c_\phi)$ as c_t, to obtain the diffusion equation for spherically symmetric flow:

$$\frac{dP}{dt}=\frac{k}{\phi\mu c_t}\frac{1}{R^2}\frac{d}{dR}\left(R^2\frac{dP}{dR}\right). \tag{A.11}$$

The solution for the "spherical point source" in an infinite reservoir, including wellbore storage effects, can be found in Brigham *et al.* (1980). The effect of wellbore skin was included in Joseph and Koederitz (1985). Spherical flow is also discussed in some detail in Stanislav and Kabir (1990).

Problem 2.1. A well with 3 in. radius is located in a 40 ft. thick reservoir that has a permeability of 30 mD and a porosity of 0.20. The total compressibility of the oil-rock system is 3×10^{-5}/psi. The initial pressure in the reservoir is 2800 psi. The well produces 448 bbl/day of oil that has a viscosity of 0.4 cP. Conversion factors can be found in Problem 1.1.

Solution: First, convert all data to SI units:

$$R_w = (3 \text{ in})\left(\frac{1 \text{ ft}}{12 \text{ in}}\right)\left(\frac{0.3048 \text{ m}}{1 \text{ ft}}\right) = 0.0762 \text{ m},$$

$$H = (40 \text{ ft})\left(\frac{0.3048 \text{ m}}{1 \text{ ft}}\right) = 12.19 \text{ m},$$

$$\mu = (0.4 \text{ cP})\left(\frac{1 \text{ P}}{100 \text{ cP}}\right)\left(\frac{0.1 \text{ Pa s}}{1 \text{ P}}\right) = 4 \times 10^{-4} \text{ Pa s},$$

$$k = (30 \text{ mD})\left(\frac{0.987 \times 10^{-15} \text{ m}^2}{1 \text{ mD}}\right) = 2.96 \times 10^{-14} \text{ m}^2,$$

$$Q = 448 \frac{\text{bbl}}{\text{day}}\left(\frac{1 \text{ day}}{24 \text{ hr}}\right)\left(\frac{1 \text{ hr}}{3600 \text{ sec}}\right)\left(\frac{0.1589 \text{ m}^3}{1 \text{ bbl}}\right)$$
$$= 8.24 \times 10^{-4}\frac{\text{m}^3}{\text{s}},$$

$$c = (3 \times 10^{-5}/\text{psi})\left(\frac{1 \text{ psi}}{6895 \text{ Pa}}\right) = 4.35 \times 10^{-9} \text{ /Pa},$$

$$P_i = (2800 \text{ psi})\left(\frac{6895 \text{ Pa}}{1 \text{ psi}}\right) = 19.31 \times 10^6 \text{ Pa}.$$

(a) How long will it take in order for the line source solution to be applicable at the wellbore wall?

According to Eq. (2.3.1), the time required for the line source solution to be applicable at the wellbore wall will be

$$t > \frac{0.25\phi\mu c R_w^2}{k}$$
$$= \frac{0.25(0.2)(0.0004 \text{ Pa s})(4.35 \times 10^{-9}/\text{Pa})(0.0762 \text{ m})^2}{(2.96 \times 10^{-14} \text{ m}^2)} = 0.017 \text{ s!} \tag{A.12}$$

So, for practical purposes, the assumption of an infinitely small borehole radius causes no problems.

(b) What is the pressure at the wellbore after six days of production, according to the line source solution?

According to the line source solution, Eq. (2.1.21),

$$P_w(t) = P_i + \frac{\mu Q}{4\pi kH} Ei\left(\frac{-\phi\mu c R_w^2}{4kt}\right). \tag{A.13}$$

First, calculate the pre-factor $\mu Q/4\pi kH$:

$$\frac{\mu Q}{4\pi kH} = \frac{(0.0004\,\text{Pa s})(8.24 \times 10^{-4}\ \text{m}^3/\text{s})}{4\pi(2.96 \times 10^{-14}\ \text{m}^2)(12.19\ \text{m})} = 7.27 \times 10^4\,\text{Pa}. \tag{A.14}$$

Now calculate the value of the variable $x = \phi\mu c R_w^2/4kt$, when $t = 6$ days $= 5.184 \times 10^5$ s:

$$\begin{aligned} x &= \frac{(0.2)(0.0004\,\text{Pa s})(4.35 \times 10^{-9}/\text{Pa})(0.0762\ \text{m})^2}{4(2.96 \times 10^{-14}\ \text{m}^2)(5.184 \times 10^5\ \text{s})} \\ &= 3.29 \times 10^{-8}. \end{aligned} \tag{A.15}$$

From Table 2.1, we find:

$$-Ei(-x) = -Ei(-3.29 \times 10^{-8}) = 16.62. \tag{A.16}$$

Combine Eqs. (A.13), (A.14) and (A.16) to get:

$$\begin{aligned} P_w(6\ \text{days}) &= 19.31 \times 10^6\,\text{Pa} - (7.27 \times 10^4\text{Pa})(16.62) \\ &= (18.10 \times 10^6\,\text{Pa})(1\ \text{psi}/6895\,\text{Pa}) = 2625\ \text{psi}. \end{aligned}$$

(c) How long will it take in order for Jacob's logarithmic approximation to be valid at the wellbore?

According to Eq. (2.4.7), the time required will be

$$t > \frac{25\phi\mu c R_w^2}{k}, \tag{A.17}$$

which is 100 times greater than the time required for the line source solution itself to be meaningful for a well having a finite radius. Hence, the logarithmic approximation will be valid at the wellbore after only 1.7 s.

(d) What is the pressure at the wellbore after six days of production, according to the logarithmic approximation?

According to Eq. (2.4.8),

$$P_w(t) = P_i + \frac{\mu Q}{4\pi kH} \ln\left(\frac{\gamma\phi\mu c R_w^2}{4kt}\right) = P_i + \frac{\mu Q}{4\pi kH} \ln(\gamma x). \quad \text{(A.18)}$$

But we know from (b) that $t = 6$ days, $x = 3.29 \times 10^{-8}$, and we also know that $\gamma = 1.781$, so

$$\begin{aligned} P_w(t) &= 19.31 \times 10^6 \text{ Pa} + (7.27 \times 10^4 \text{ Pa}) \ln[1.781(3.29 \times 10^{-8})] \\ &= 19.31 \times 10^6 \text{ Pa} - 1.21 \times 10^6 \text{ Pa} \\ &= 18.10 \times 10^6 \text{ Pa} = 2625 \text{ psi}. \end{aligned}$$

As expected, based on the answer to part (c), the logarithmic approximation gives the correct pressure at the wellbore.

(e) Answer questions (b)–(d) for a location that is 800 ft. (horizontally) away from the wellbore.

Using the same equations as in parts (a)–(d), we find that:

$$P(R = 800 \text{ ft}, \ t = 6 \text{ days, exact solution}) = 2791 \text{ psi},$$

$$P(R = 800 \text{ ft}, \ t = 6 \text{ days, logarithmic approximation}) = 2795 \text{ psi}.$$

So, the logarithmic approximation gives a drawdown of 5 psi, whereas the actual drawdown is 9 psi! This is consistent with the fact that, according to Eq. (2.4.7), the logarithmic approximation will not be accurate at a radius of 800 ft. until 207 days have elapsed.

Problem 2.2. A well with a radius of 0.3 ft. produces 200 bbl/day of oil, with viscosity 0.6 cP, from a 20 ft. thick reservoir. The wellbore pressures are as follows:

t (mins)	0	5	10	20	60	120	480	1440	2880	5760
P_w (psi)	4000	3943	3938	3933	3926	3921	3911	3904	3899	3894

Estimate the permeability and the storativity of the reservoir, using the semi-log method presented in Section 2.6.

Solution: First, plot the well pressure versus the logarithm of time; it is not necessary to convert the data to SI units (see Figure A.1).

Next, look for a straight line at *late* times, and find its slope:

$$m = \left|\frac{\Delta P}{\Delta \ln t}\right| = \frac{4004 - 3890}{7(2.303)} = 7.07 \text{ psi } \times \left(\frac{6895\,\text{Pa}}{\text{psi}}\right) = 48758\,\text{Pa}.$$

Note 1: 4004 psi is the value obtained by extrapolating the straight line back to $t = 0.001$ min. We use this time for convenience, so that Δt covers an integral number of log cycles; this time has no physical significance.

Note 2: $\Delta \ln t$ has the same numerical value *regardless* of which units are used for t! In this case, Δt = "seven orders of magnitude", so $\Delta \ln t = 7(\ln 10) = 7(2.303)$.

Now, calculate k from Eq. (2.6.4). To do this, first convert the data to SI units:

$$\mu = 0.6 \text{ cP} \times \left(\frac{0.001\,\text{Pa s}}{\text{cP}}\right) = 0.0006\,\text{Pa s},$$

$$Q = 200\frac{\text{bbl}}{\text{day}} \times \frac{0.1589 \text{ m}^3}{\text{bbl}} \times \frac{\text{day}}{24 \text{ hr}} \times \frac{\text{hr}}{3600 \text{ s}} = 3.68 \times 10^{-4}\frac{\text{m}^3}{\text{s}},$$

$$H = 20 \text{ ft} \times (0.3048 \text{ m/ft}) = 6.096 \text{ m},$$

$$k = \frac{\mu Q}{4\pi m H} = \frac{(0.0006\,\text{Pa s})(3.68 \times 10^{-4} \text{ m}^3/\text{s})}{4\pi(48,758\,\text{Pa})(6.096 \text{ m})}$$

$$= 5.91 \times 10^{-14} \text{ m}^2 \left(\frac{1 \text{ mD}}{0.987 \times 10^{-15} \text{ m}^2}\right) = 59.9 \text{ mD}.$$

Note that the "exact value", i.e. the value used to generate the synthetic pressure data, was 60 mD.

To estimate the storativity term, ϕc, extrapolate the straight line back to the initial pressure, 4000 psi (see Figure A.1). The extrapolated line crosses the horizontal line $P_w = 4000$ psi at $t^* = 0.0016$ min $= 0.096$ sec. We then use Eq. (2.6.5) to calculate ϕc.

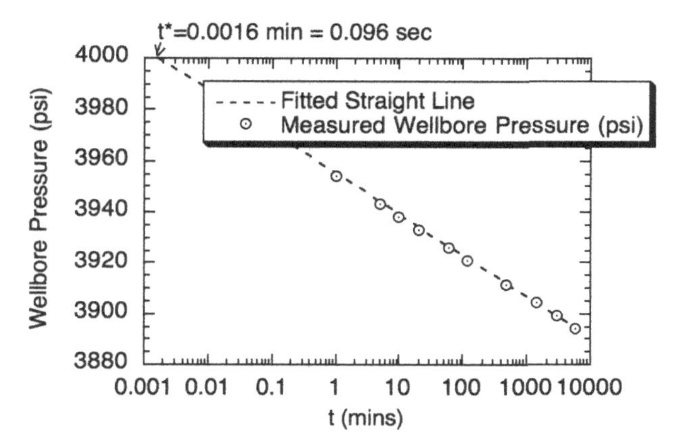

Figure A.1. Semi-log plot of pressure used in the estimation of reservoir properties.

First, we need to convert the wellbore radius to SI units: $R_w = (0.3\,\text{ft}) \times (0.3048\,\text{m/ft}) = 0.0914\,\text{m}$. Finally, using Eq. (2.6.5), we find

$$\phi c = \frac{2.246kt^*}{\mu R_w^2} = \frac{2.246(5.91 \times 10^{-14}\ \text{m}^2)(0.096\ \text{s})}{(0.0006\,\text{Pa s})(0.0914\ \text{m})^2}$$
$$= 2.54 \times 10^{-9}\,/\text{Pa}.$$

Problem 3.1. Which, if any, of the following differential equations are linear, and why (or why not)?

(a) $\dfrac{d^2y}{dx^2} + y\dfrac{dy}{dx} + y = 0.$

(b) $\dfrac{d^2y}{dx^2} + x\dfrac{dy}{dx} + y = 0.$

(c) $\dfrac{d^2y}{dx^2} + x\dfrac{dy}{dx} + xy = 0.$

Solution: The main test to see if an equation is linear is that a linear differential equation can only contain the dependent variable (in this case, y), or its derivatives, to the first power. If the equation contains y or any of its derivatives taken to a higher power, or multiplied together, then it is nonlinear. So, terms such as $y(dy/dx)$ will make the equation nonlinear.

By this rule, it seems that Eq. (a) is nonlinear, but Eqs. (b) and (c) are linear. Note that multiplying y by the independent variable x is OK — this will not make the equation nonlinear. However, just to be sure, we should also check the basic definition of linearity. So, let us denote the "operator" in Eq. (a) by "$M(y)$", then plug $y = y_1 + y_2$ into $M(y)$, and see what we get:

$$\begin{aligned} M(y_1 + y_2) &= \frac{d^2(y_1 + y_2)}{dx^2} + (y_1 + y_2)\frac{d(y_1 + y_2)}{dx} + (y_1 + y_2) \\ &= \frac{d^2 y_1}{dx^2} + \frac{d^2 y_2}{dx^2} + (y_1 + y_2)\left(\frac{dy_1}{dx} + \frac{dy_2}{dx}\right) + (y_1 + y_2) \\ &= \frac{d^2 y_1}{dx^2} + \frac{d^2 y_2}{dx^2} + y_1\frac{dy_1}{dx} + y_2\frac{dy_1}{dx} + y_1\frac{dy_2}{dx} \\ &\quad + y_2\frac{dy_2}{dx} + y_1 + y_2 \\ &= \left(\frac{d^2 y_1}{dx^2} + y_1\frac{dy_1}{dx} + y_1\right) + \left(\frac{d^2 y_2}{dx^2} + y_2\frac{dy_2}{dx} + y_2\right) \\ &\quad + y_1\frac{dy_2}{dx} + y_2\frac{dy_1}{dx} \\ &= M(y_1) + M(y_2) + y_1\frac{dy_2}{dx} + y_2\frac{dy_1}{dx}. \end{aligned}$$

Because of the terms such as $y_1(dy_2/dx)$, in general it will *not* be true that $M(y_1 + y_2) = M(y_1) + M(y_2)$ for arbitrary y_1 and y_2; therefore, Eq. (a) is nonlinear. Similarly, we can show that Eqs. (b) and (c) *are* linear.

Additionally, to rigorously verify linearity, we must also check that $M(cy) = cM(y)$ for any function y and any constant c. It is easy to see that Eq. (a) also fails this test, but (b) and (c) satisfy this criterion.

The property of linearity is crucial, because all the standard methods used to solve the diffusion equation (Laplace transforms, eigenfunction expansions, Green's functions, etc.) work only for *linear* equations.

Problem 3.2. Find an expression for the wellbore pressure in a vertical well drilled into an infinite reservoir, if the production rate increases linearly as a function of time according to $Q(t) = Q_* t/t_*$, where Q_* and t_* are constants. Use convolution, in the form of either Eq. (3.4.6) or Eq. (3.4.10), and recall that $\Delta P_Q(R,t)$ for a well in an infinite reservoir is given by Eq. (3.3.6).

Solution: According to Eq. (3.4.10), the drawdown due to an *arbitrary* flow rate $Q(t)$ can be found from the solution for *constant* flow rate by evaluating the following convolution integral:

$$\Delta P(R,t) = \int_0^t Q(\tau) \frac{d\Delta P_Q(R,t-\tau)}{dt} d\tau. \tag{A.19}$$

where $\Delta P_Q(R,t)$, given by Eq. (3.3.6), is the drawdown per unit flow rate for the case of constant flow rate:

$$\begin{aligned}\Delta P_Q(R,t) &= \frac{-\mu}{4\pi kH} Ei\left(\frac{-\phi\mu cR^2}{4kt}\right) \\ &= \frac{\mu}{4\pi kH} \int_{\frac{\phi\mu cR^2}{4kt}}^{\infty} \frac{e^{-u}}{u} du. \end{aligned} \tag{A.20}$$

Using the chain rule on the term on the right side of Eq. (A.20), the derivative appearing in Eq. (A.19) is found to be

$$\begin{aligned}\frac{d\Delta P_Q(R,t)}{dt} &= \frac{-\mu}{4\pi kH} \left[\frac{\exp(-\phi\mu cR^2/4kt)}{(\phi\mu cR^2/4kt)}\right] \frac{d(\phi\mu cR^2/4kt)}{dt} \\ &= \frac{\mu}{4\pi kHt} \exp^{-(\phi\mu cR^2/4kt)}. \end{aligned} \tag{A.21}$$

Using $Q(t) = Q_* t/t_*$ and Eq. (A.21) in Eq. (A.19), we find

$$\Delta P(R,t) = \int_0^{\infty} \frac{Q_* \tau\mu}{4\pi kHt_*} \frac{\exp^{-[\phi\mu cR^2/4k(t-\tau)]}}{(t-\tau)} d\tau. \tag{A.22}$$

We now make the following change of variables:

$$x = \frac{\phi\mu cR^2}{4k(t-\tau)} \quad \rightarrow \quad \tau = t - \frac{\phi\mu cR^2}{4kx} \quad \rightarrow \quad d\tau = \frac{\phi\mu cR^2}{4kx^2} dx. \tag{A.23}$$

Note that τ is the integration variable in Eq. (A.22), whereas t is just a parameter. We also need to change the limits of integration:

$$\text{when } \tau = 0, \quad x = \frac{\phi\mu c R^2}{kt}, \tag{A.24}$$

$$\text{when } \tau = t, \quad x = \infty. \tag{A.25}$$

Inserting Eqs. (A.23–A.25) into Eq. (A.22) gives

$$\begin{aligned}\Delta P(R,t) &= \frac{Q_*\mu}{4\pi kHt_*}\int_{\frac{\phi\mu cR^2}{4kt}}^{\infty}\left(t - \frac{\phi\mu cR^2}{4kx}\right)\frac{e^{-x}}{x}dx \\ &= \frac{Q_*\mu t}{4\pi kHt_*}\int_{\frac{\phi\mu cR^2}{4kt}}^{\infty}\frac{e^{-x}}{x}dx \\ &\quad - \frac{Q_*\mu t}{4\pi kHt_*}\left(\frac{\phi\mu cR^2}{4kt}\right)\int_{\frac{\phi\mu cR^2}{4kt}}^{\infty}\frac{e^{-x}}{x^2}dx.\end{aligned} \tag{A.26}$$

The first integral on the right side in Eq. (A.26) is, aside from the sign, the Ei function, and noting that $Q(t) = Q_*t/t_*$, thus far we have:

$$\begin{aligned}\Delta P(R,t) &= \frac{-\mu Q(t)}{4\pi kH}Ei\left(\frac{-\phi\mu cR^2}{4kt}\right) \\ &\quad - \frac{\mu Q(t)}{4\pi kH}\left(\frac{\phi\mu cR^2}{4kt}\right)\int_{\frac{\phi\mu cR^2}{4kt}}^{\infty}\frac{e^{-x}}{x^2}dx.\end{aligned} \tag{A.27}$$

To evaluate the remaining integral, we use integration-by-parts, with the following choices for "u" and "v":

$$u = e^{-x}, \quad dv = \frac{-1}{x^2}dx, \quad du = -e^{-x}dx, \quad v = \frac{1}{x}. \tag{A.28}$$

The integral in Eq. (A.27) can now be evaluated as follows:

$$\int_w^{\infty}\frac{-e^{-x}}{x^2}dx = \left.\frac{e^{-x}}{x}\right]_w^{\infty} + \int_w^{\infty}\frac{e^{-x}}{x}dx = -\frac{e^{-w}}{w} - Ei(-w). \tag{A.29}$$

Using result (A.29) in Eq. (A.27) gives

$$\Delta P(R,t) = \frac{-\mu Q(t)}{4\pi kH} Ei\left(\frac{-\phi\mu cR^2}{4kt}\right)$$

$$- \frac{\mu Q(t)}{4\pi kH}\left(\frac{\phi\mu cR^2}{4kt}\right)\left[\frac{e^{\left(\frac{-\phi\mu cR^2}{4kt}\right)}}{\left(\frac{\phi\mu cR^2}{4kt}\right)} + Ei\left(\frac{-\phi\mu cR^2}{4kt}\right)\right]. \tag{A.30}$$

This result can be written in dimensionless form as

$$\Delta P_D(R,t) = -\frac{1}{2}\left[\left(1+\frac{1}{4t_D}\right)Ei(-1/4t_D) + e^{-1/4t_D}\right], \tag{A.31}$$

where the dimensionless time is defined in the usual way as

$$t_D = \frac{kt}{\phi\mu cR^2}, \tag{A.32}$$

and the dimensionless drawdown is defined as

$$\Delta P_D(R,t) = \frac{2\pi kH\Delta P(t)}{\mu Q(t)}. \tag{A.33}$$

When $t_D > 50$, the exponential approximation can be used for the Ei function, the term $e^{-1/4t_D}$ is essentially equal to 1, and the term $1/4t_D$ is negligible compared to 1, in which case Eq. (A.31) reduces to

$$\Delta P_D(R,t) = -\frac{1}{2}\left[\ln(1/4t_D) + \ln\gamma + 1\right] = \frac{1}{2}\left[\ln(4t_D) - \ln\gamma - 1\right]$$

$$= \frac{1}{2}(\ln 4 + \ln t_D - \ln\gamma - 1) = \frac{1}{2}(\ln t_D - 0.191). \tag{A.34}$$

The graph of the dimensionless drawdown versus dimensionless time is shown in Figure A.2, with the exact solution shown by the solid curve, and the late-time approximation shown by the dashed curve. The dotted vertical line at $t_D = 50$ indicates the time at which the approximate solution becomes accurate to within 1%.

Dimensionless drawdown versus dimensionless time, as given by the exact expression, Eq. (A.31), and the approximate expression, Eq. (A.34).

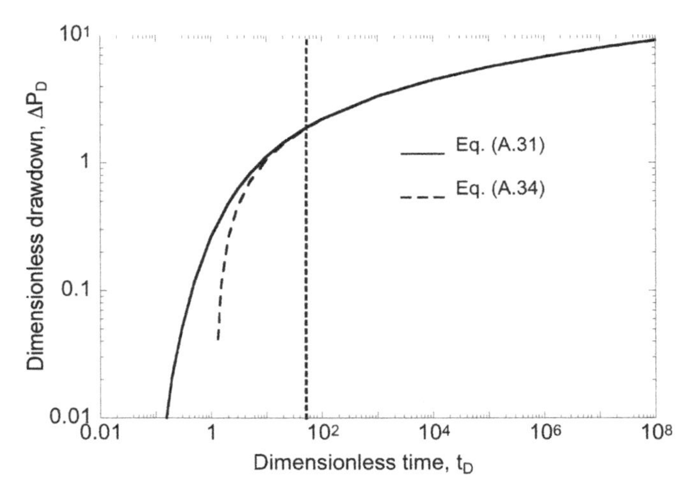

Figure A.2. Dimensionless drawdown versus Dimensionless time, as given by the exact expression, Eq. (A.31), and the approximate expression, Eq. (A.34).

Problem 4.1. As explained in Section 4.2, a doubling of the slope on a semi-log plot of drawdown versus time indicates the presence of an impermeable linear fault. The drawdown data can also be used to find the distance from the well to the fault, as follows. If we plot the data and then fit two straight lines through the early-time and late-time data, the time at which these lines intersect is called t'_{Dw}. Show that the distance to the fault is then given by the equation $d = (0.5615t'_{\mathrm{Dw}})^{1/2}R_w$.

Solution: According to Section 4.2, in the time regime defined by

$$25 < t_{\mathrm{Dw}} < 0.3(d/R_w)^2, \tag{A.35}$$

the drawdown is essentially that of a well in an infinite reservoir, i.e. Eq. (4.2.7):

$$P_w(t) = P_i - \frac{\mu Q}{4\pi kH}\ln\left(\frac{2.246kt}{\phi\mu c R_w^2}\right). \tag{A.36}$$

In the time regime defined by

$$t_{\mathrm{Dw}} > 100(d/R_w)^2, \tag{A.37}$$

the drawdown in the production well is described by Eq. (4.2.10)

$$P_w(t) = P_i - \frac{\mu Q}{4\pi kH}\ln\left(\frac{2.246kt}{\phi\mu c R_w^2}\right) - \frac{\mu Q}{4\pi kH}\ln\left[\frac{2.246kt}{\phi\mu c(2d)^2}\right]. \tag{A.38}$$

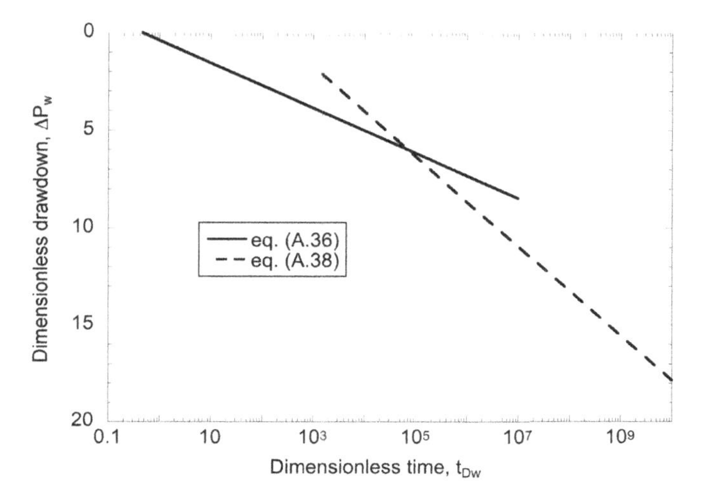

Figure A.3. Intersection of the two lines given by Eqs. (A.36) and (A.38).

Both Eqs. (A.36) and (A.38) yield straight lines on a semi-log plot of P_w versus $\ln t$ (see Figure A.3). The time at which these two lines cross is found by setting the two pressures, from Eqs. (A.36) and (A.38), equal to each other. They will only be equal if the third term in Eq. (A.38) is zero, i.e. if

$$\frac{\mu Q}{4\pi kH} \ln \left[\frac{2.246kt}{\phi\mu c(2d)^2} \right] = 0. \tag{A.39}$$

But $\ln x$ will be zero only when $x = 1$, which occurs when

$$2.246kt = \phi\mu c(2d)^2 \quad \rightarrow \quad d^2 = \frac{0.5615kt}{\phi\mu c}. \tag{A.40}$$

In dimensionless form, this can be written as

$$\frac{d^2}{R_w^2} = \frac{0.5615kt'}{\phi\mu c R_w^2} = 0.5615t'_{\mathrm{Dw}} \quad \rightarrow \quad d = 0.749(t'_{\mathrm{Dw}})^{1/2} R_w. \tag{A.41}$$

Problem 4.2. The curves in Figure 4.5 were drawn for the case $d = 200R_w$. What would the curves look like for the case of a fault located at a distance $d = 400R_w$?

Solution: Based on the discussion given in the solution to Problem 4.1, we see that at early times, the drawdown will still be given by Eq. (A.36). At a time defined by Eq. (A.41), the pressure will veer off along another straight line with twice the slope of the original line. All other factors being equal, the time at which the second line diverges from the original line will scale according to d^2.

Hence, if d increases by a factor of 2, the time at which the second curve veers off from the first will be increased by a factor of 4. But the time axis is plotted logarithmically, so this has the effect of shifting the second curve to the *right* by an amount $\log_{10}(4)$, as in Figure A.4.

Problem 4.3. Consider a well-located equidistant from two orthogonal boundaries, as in Figure 4.3, but imagine that the boundaries are *constant-pressure* boundaries, rather than impermeable boundaries. How would you utilise the method of images to find the drawdown in this well?

Solution: Recall that to account for one constant-pressure linear boundary, we positioned an image well symmetrically across the boundary, and we made it an injection well.

This method works because the drawdown caused by image well 1 along the vertical line in Figure A.5 is equal in magnitude, but

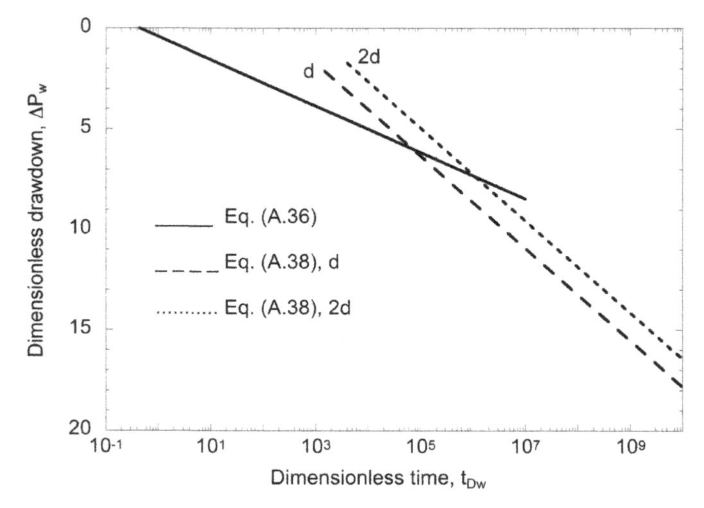

Figure A.4. If the distance to the fault is doubled, the drawdown curve is shifted to the right by $\log_{10}(4)$.

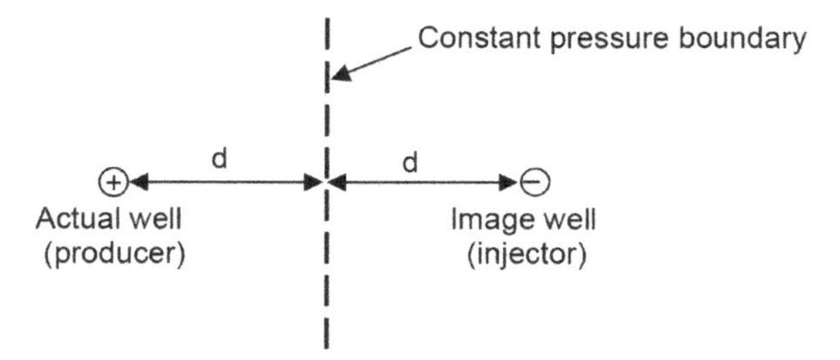

Figure A.5. Image well (injector) required in order to cause the vertical dotted line to become a constant-pressure boundary.

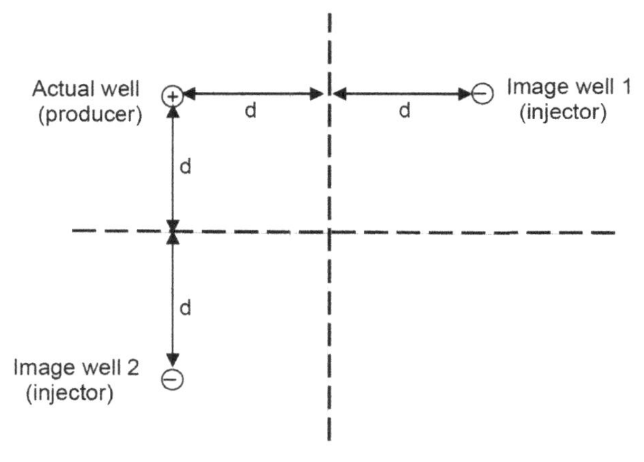

Figure A.6. Actual production well and two image (injection) wells. Image well 1 is located so as to cancel out the drawdown due to the actual well along the vertical dotted line; image well 2 does the same for the horizontal line.

opposite in sign, to that caused by the actual production well. So, if we have two constant-pressure boundaries intersecting at right angles, we might try to install two image wells as injectors (−), one located symmetrically to the actual well (+) with respect to the vertical boundary, and one located symmetrically to the actual well with respect to the horizontal boundary, as in Figure A.6.

However, this is still not quite correct, because image well 2 will cause a drawdown along the vertical boundary that is not compensated for by either the actual well or the first image well.

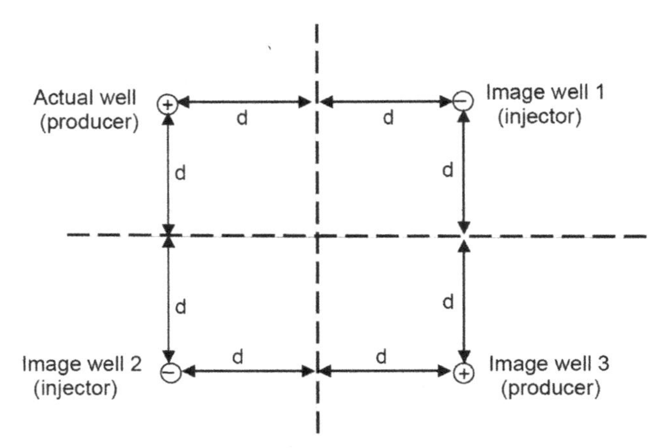

Figure A.7. Inclusion of a third image well, to cancel out the unwanted drawdown due to well 2 along the vertical dotted line, and to cancel out the drawdown due to well 1 along the horizontal dotted line.

We therefore need another well to cancel out the drawdown caused by image well 2 along the vertical boundary. This can be achieved by placing another *production* well (well 3, +) in a position symmetric to well 2 with respect to the vertical boundary, as in Figure A.7. Note that image well 3 will also cancel out the drawdown that well 1 causes along the horizontal boundary.

To check that this configuration is correct, let us examine the drawdown due to these four wells. Consider a generic point in the reservoir, located at distance R from the actual well, distance R_1 from well 1, etc., as in Figure A.8.

From inspection of Figure A.8, the total drawdown can be written as

$$\frac{4\pi kH[P(R,t)-P_i]}{\mu Q} = Ei\left(\frac{-\phi\mu cR^2}{4kt}\right) + Ei\left(\frac{-\phi\mu cR_3^2}{4kt}\right) - Ei\left(\frac{-\phi\mu cR_1^2}{4kt}\right) - Ei\left(\frac{-\phi\mu cR_2^2}{4kt}\right). \quad (A.42)$$

Now consider a point that lies on the horizontal constant-pressure boundary, such as in Figure A.9.

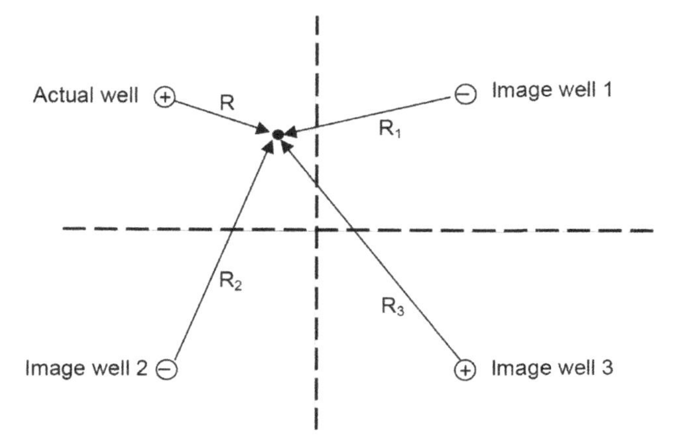

Figure A.8. Image wells and distances to a generic point in the reservoir.

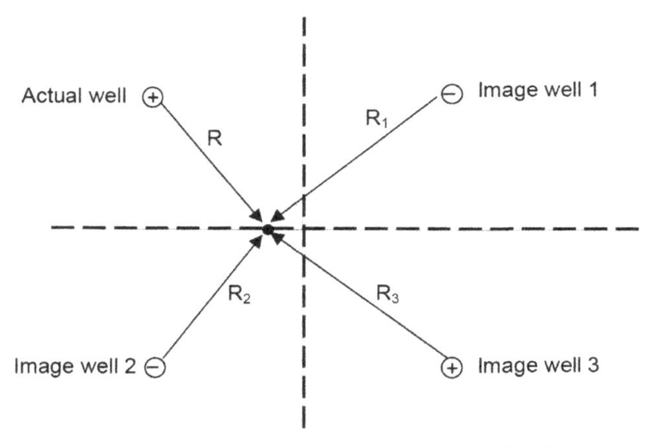

Figure A.9. Radius variables for a point located on the horizontal constant-pressure boundary.

Note that $R_2 = R$ and $R_3 = R_1$, so from Eq. (A.42) the total drawdown at this point can be written as

$$\frac{4\pi kH[P(R,t) - P_i]}{\mu Q} = Ei\left(\frac{-\phi\mu cR^2}{4kt}\right) + Ei\left(\frac{-\phi\mu cR_1^2}{4kt}\right) - Ei\left(\frac{-\phi\mu cR_1^2}{4kt}\right) - Ei\left(\frac{-\phi\mu cR^2}{4kt}\right) = 0, \tag{A.43}$$

which proves that the pressure along the horizontal boundary remains at P_i at all times.

The same is true for a point along the vertical dotted line, thus proving that we have constructed the solution for a well symmetrically bounded by two orthogonal, intersecting constant-pressure boundaries.

Finally, the drawdown in the actual well is found from Eq. (A.42) by setting $R = R_w$, $R_1 = R_2 = 2d$, and $R_3 = 2\sqrt{2}d$, to get

$$\frac{4\pi kH[P_w(t) - P_i]}{\mu Q} = Ei\left(\frac{-\phi\mu c R_w^2}{4kt}\right) + Ei\left(\frac{-2\phi\mu c d^2}{kt}\right) - 2Ei\left(\frac{-\phi\mu c d^2}{kt}\right). \tag{A.44}$$

Problem 5.1. Imagine a wellbore that is filled with liquid only up to some height h above the top of the reservoir. The liquid has density ρ. Even if this liquid were incompressible, a wellbore storage effect would still occur, due to the raising or lowering of the fluid column. Derive an expression for the wellbore storage coefficient C_s in this case.

Solution: The basic definition of the wellbore storage coefficient is embodied in Eq. (5.3.2):

$$Q_{\text{sf}} - Q_{\text{wh}} = C_s \frac{dP_w}{dt}. \tag{A.45}$$

Consider an increment of time Δt. The net flow of fluid into the well during this time period will be $(Q_{\text{sf}} - Q_{\text{wh}})\Delta t$. This must equal the change in the volume of fluid stored in the well, which is $A\Delta h$, where A is the cross-sectional area of the well. Inserting these two expressions into Eq. (A.45) gives

$$Q_{\text{sf}} - Q_{\text{wh}} = A \frac{dh}{dt}. \tag{A.46}$$

Equating the two expressions (A.45) and (A.46) shows that $C_s(dP_w/dt) = A(dh/dt)$. The pressure in the liquid at the top of the reservoir will be ρgh, so $dP_w/dt = \rho g(dh/dt)$, which shows that $C_s \rho g = A$, which in turn implies that $C_s = A/\rho g$.

Problem 6.1. Starting with the expression for the pressure drawdown, Eq. (6.4.6), show that the transition regime for a closed circular reservoir with constant production rate ends when $t \approx 0.3\phi\mu c R_e^2/k$.

Hints:

(a) $e^{-x} \approx 0$ when $x > 4$, so all terms in the series will be negligible when $\lambda_n^2 t_D > 4$ for all n.
(b) When R_{De} is large, the *first* eigenvalue, λ_1, defined as the smallest value of λ that satisfies Eq. (6.4.4), is *very* small. This fact should help you to estimate the value of λ_1 as a function of R_{De}, by making the reasonable assumption that λ_1 is inversely proportional to R_{De}.
(c) Make use of Eqs. (6.2.37), (6.2.39) and (6.2.48) and Figure 6.3.

Solution: Recall from Section (6.4) that the wellbore pressure for a constant flow rate well in a closed circular reservoir is given by

$$\Delta P_{\text{Dw}}(t_D) = \frac{2t_D}{R_{\text{De}}^2} + \ln R_{\text{De}} - \frac{3}{4} + \sum_{n=1}^{\infty} \frac{2J_1^2(\lambda_n R_{\text{De}})e^{-\lambda_n^2 t_D}}{\lambda_n^2[J_1^2(\lambda_n R_{\text{De}}) - J_1^2(\lambda_n)]}, \tag{A.47}$$

in which the eigenvalues λ_n are defined, implicitly, as the roots of the following equation:

$$J_1(\lambda_n)Y_1(\lambda_n R_{\text{De}}) - Y_1(\lambda_n)J_1(\lambda_n R_{\text{De}}) = 0, \tag{A.48}$$

and the dimensionless variables are defined as in Section 6.3.

The transition regime ends when all of the exponential terms in Eq. (A.47) are negligible, which we can define to occur when the values of the exponentials are all less than 0.01. Since $e^{-4.6} = 0.01$, we can say that an exponential term $e^{-x} \approx 0$ will be "negligible" when the variable x is greater than 4.6.

Recall from Eq. (6.2.48) that there will always be an infinite number of positive eigenvalues λ_n, such that

$$0 < \lambda_1 < \lambda_2 < \lambda_3 < \cdots < \lambda_n < \cdots. \tag{A.49}$$

Equation (A.49) implies that

$$0 < \lambda_1^2 t_D < \lambda_2^2 t_D < \lambda_3^2 t_D < \cdots < \lambda_n^2 t_D < \cdots, \tag{A.50}$$

which in turn implies that

$$e^{-\lambda_1^2 t_D} > e^{-\lambda_2^2 t_D} > e^{-\lambda_3^2 t_D} > \cdots > e^{-\lambda_n^2 t_D} > \cdots . \tag{A.51}$$

Hence, if we want each exponential term to be less than 0.01, we only need to check that the first term is < 0.01. This will occur when

$$\lambda_1^2 t_D > 4.6 \quad \rightarrow \quad t_D > \frac{4.6}{\lambda_1^2}. \tag{A.52}$$

So, our main task is to estimate the value of λ_1, the smallest eigenvalue, which is the smallest value of λ that satisfies Eq. (A.48):

$$J_1(\lambda_n)Y_1(\lambda_n R_{\text{De}}) - Y_1(\lambda_n)J_1(\lambda_n R_{\text{De}}) = 0. \tag{A.53}$$

Physically, we expect that the time required for the transition period to end will be larger for larger reservoirs, and vice versa. So, Eq. (A.52) implies that we should expect that as R_{De} increases, λ_1 will decrease, and, in the limit,

$$\lambda_1 \rightarrow 0 \quad \text{as} \quad R_{\text{De}} \rightarrow \infty. \tag{A.54}$$

Since relevant values of R_{De} are usually >100, we know that the root of Eq. (A.53) that we are looking for will be very close to zero. Let us now make the reasonable assumption that λ_1 and R_{De} are inversely related, i.e. $\lambda_1 = c/R_{\text{De}}$, for some constant c, whose value we do not yet know.

We must now examine all the terms in Eq. (A.48), and see if we can simplify them when λ_1 is very small. First, recall from Eq. (6.3.5) that

$$J_1(x) \equiv -\frac{dJ_0(x)}{dx}, \quad Y_1(x) \equiv -\frac{dY_0(x)}{dx}. \tag{A.55}$$

Next, recall from Eq. (6.2.37) that for small values of x,

$$J_0(x) = 1 - \frac{x^2}{4} + \cdots . \tag{A.56}$$

Combining Eqs. (A.55) and (A.56) shows that, for small x,

$$J_1(x) = \frac{-dJ_0(x)}{dx} = \frac{x}{2} + \cdots . \tag{A.57}$$

Next, recall the definition of $Y_0(x)$ given by Eq. (6.2.39)

$$Y_0(x) = \frac{2}{\pi}\ln(\gamma x/2)J_0(x) - \frac{2}{\pi}\sum_{n=1}^{\infty}\frac{(-1)^n h_n}{(n!)^2 2^{2n}}x^{2n}. \tag{A.58}$$

For small x, all the terms in the series will be very small. Also, we know from Eq. (A.56) that $J_0(x) \approx 1$ for small x. Hence, we see from Eq. (A.58) that, for small x,

$$Y_0(x) \approx \frac{2}{\pi}\ln(\gamma x/2) = \frac{2}{\pi}(\ln x + \ln\gamma - \ln 2) \approx \frac{2}{\pi}\ln x, \tag{A.59}$$

where we retain only the $\ln x$ term, because this is the only term that becomes large in magnitude as x becomes small. Now, using the second equation in Eq. (A.55), we have

$$Y_1(x) = -\frac{dY_0(x)}{dx} \approx -\frac{d}{dx}\left(\frac{2}{\pi}\ln x\right) = \frac{-2}{\pi x}. \tag{A.60}$$

With our assumption that $\lambda_1 = c/R_{\mathrm{De}}$, Eq. (A.48) takes the form

$$J_1(\lambda_1)Y_1(c) - Y_1(\lambda_1)J_1(c) = 0. \tag{A.61}$$

Now, λ_1 is small, and c is some (albeit unknown) finite number, so we can use Eq. (A.57) for $J_1(\lambda_1)$, and Eq. (A.60) for $Y_1(\lambda_1)$, to find

$$\frac{\lambda_1}{2}Y_1(c) + \frac{2}{\pi\lambda_1}J_1(c) = 0,$$

$$\rightarrow \quad J_1(c) = -\frac{\pi\lambda_1^2}{4}Y_1(c) = -\frac{\pi c^2}{4R_{\mathrm{De}}^2}Y_1(c). \tag{A.62}$$

As R_{De} gets very large, which in practice will always will be the case, the right-hand side of Eq. (A.62) goes to zero. Hence, the left-hand side must also go to zero. So, the value of c that we are looking for is the smallest value that satisfies the equation

$$J_1(c) = 0. \tag{A.63}$$

This number is called "the first zero of J_1", and can be found in most books on advanced applied mathematics. Or, we can be clever and recall that, by Eq. (A.55), $J_1(c)$ will be zero when the derivative of

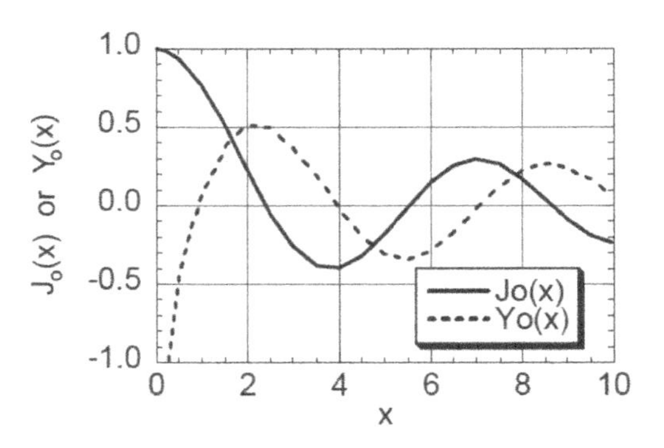

Figure A.10. Bessel functions of order zero.

$J_0(c)$ is zero — and we can find this value of c from Figure 6.2.2, which is repeated in Figure A.10.

We see from Figure A.10 that the smallest value of c for which the derivative $J_0(c)$ is zero is roughly $c = 3.8$; the exact value is actually 3.83. Recalling that $\lambda_1 = c/R_{\text{De}}$, we have

$$\lambda_1 \approx \frac{3.8}{R_{\text{De}}}. \tag{A.64}$$

So, criterion (A.52) for the end of the transition regime becomes

$$t_D > \frac{4.6}{(3.8/R_{\text{De}})^2} \approx 0.3R_{\text{De}}^2. \tag{A.65}$$

If we use Eqs. (6.2.5) and (6.2.6) to express this result in terms of actual physical variables, we find that the transition regime ends when $t \approx 0.3\phi\mu c R_e^2/k$. Hence, although the end of the transition regime seemed to depend on the wellbore radius, through the dependence of R_{De} on R_w, when Eq. (A.65) is expressed in terms of actual physical (dimensional) variables, it becomes clear that this time depends on the overall size of the reservoir, but *not* on the radius of the wellbore. This example shows that although it is convenient to use dimensionless parameters during the solution process, and when plotting the results, more physical insight can always be gained by eventually reverting to physical variables.

Problem 6.2. Starting with Eq. (6.4.2), calculate the average pressure in the reservoir during the finite reservoir regime, $t_{\text{Dw}} \equiv$

$t_D > 0.3R_{\mathrm{De}}^2$, as a function of time. Is your result consistent with the mass-balance embodied in Eq. (6.4.16)?

Solution: Recall Eq. (6.4.2) for the pressure distribution in the reservoir:

$$\Delta P_D(R_D, t_D) = \frac{1}{R_{\mathrm{De}}^2 - 1}\left[\frac{R_D^2}{2} + 2t_D - R_{\mathrm{De}}^2 \ln R_D\right] - \left[\frac{3R_{\mathrm{De}}^4 - 4R_{\mathrm{De}}^4 \ln R_{\mathrm{De}} - 2R_{\mathrm{De}}^2 - 1}{4(R_{\mathrm{De}}^2 - 1)^2}\right] + \sum_{n=1}^{\infty} \frac{\pi J_1^2(\lambda_n R_{\mathrm{De}}) U_n(\lambda_n R_D)}{\lambda_n [J_1^2(\lambda_n R_{\mathrm{De}}) - J_1^2(\lambda_n)]} e^{-\lambda_n^2 t_D}. \quad \text{(A.66)}$$

When $t_D > 0.3R_{\mathrm{De}}^2$, the terms in the series have died out, and the pressure is given by

$$\Delta P_D(R_D, t_D) = \frac{1}{R_{\mathrm{De}}^2 - 1}\left[\frac{R_D^2}{2} + 2t_D - R_{\mathrm{De}}^2 \ln R_D\right] - \left[\frac{3R_{\mathrm{De}}^4 - 4R_{\mathrm{De}}^4 \ln R_{\mathrm{De}} - 2R_{\mathrm{De}}^2 - 1}{4(R_{\mathrm{De}}^2 - 1)^2}\right]. \quad \text{(A.67)}$$

Since in practice it is always the case that $R_{\mathrm{De}}^2 \gg 1$, we can simplify Eq. (A.67) to the form

$$\Delta P_D(R_D, t_D) = \frac{R_D^2}{2R_{\mathrm{De}}^2} + \frac{2t_D}{R_{\mathrm{De}}^2} - \ln R_D - \frac{3}{4} + \ln R_{\mathrm{De}}. \quad \text{(A.68)}$$

This expression is difficult to interpret physically, so we will revert to dimensional variables:

$$P(R,t) = P_i - \frac{\mu Q}{2\pi kH}\left[\frac{R^2}{2R_e^2} + \frac{2kt}{\phi\mu c R_e^2} - \ln\left(\frac{R}{R_w}\right) - \frac{3}{4} + \ln\left(\frac{R_e}{R_w}\right)\right],$$
$$= P_i - \frac{\mu Q}{2\pi kH}\left[\frac{R^2}{2R_e^2} + \frac{2kt}{\phi\mu c R_e^2} - \ln\left(\frac{R}{R_e}\right) - \frac{3}{4}\right]. \quad \text{(A.69)}$$

To find the average pressure, we must integrate the pressure throughout the entire reservoir, and divide by the volume of the reservoir;

i.e. in general,

$$\langle P(t)\rangle = \frac{1}{V}\int P(R,t)dV = \frac{1}{AH}\int P(R,t)HdA = \frac{1}{A}\int P(R,t)dA. \tag{A.70}$$

The second, fourth and fifth terms inside the brackets in Eq. (A.69) are constant, so they do not need to be integrated, as the average value of a constant P_o is simply P_o.

To integrate the terms that vary with R, we note that $dA = 2\pi RdR$, and so, for the R-dependent terms

$$\langle P(R,t)\rangle = \frac{2\pi}{A}\int P(R,t)RdR. \tag{A.71}$$

Integrating from R_w to R_e, the average value of R^2 can be calculated to be

$$\langle R^2\rangle = \frac{2\pi}{A}\int R^3 dR = \frac{\pi}{2A}\left[R_e^4 - R_w^4\right] \approx \frac{\pi R_e^4}{2A}. \tag{A.72}$$

But the area of the reservoir is given by

$$A = \pi\left[R_e^2 - R_w^2\right] \approx \pi R_e^2, \tag{A.73}$$

so the average value of R^2 is

$$\langle R^2\rangle \approx \frac{\pi R_e^4}{2\pi R_e^2} = \frac{R_e^2}{2}. \tag{A.74}$$

Similarly, after integrating by parts and using the fact that $R_e \gg R_w$, the average value of the term $\ln R$ can be found to be given by

$$\langle \ln R\rangle \approx \ln R_e - \frac{1}{2}. \tag{A.75}$$

Inserting Eqs. (A.69), (A.74) and (A.75) into Eq. (A.70), we find

$$\begin{aligned}\langle P(R,t)\rangle &= P_i - \frac{\mu Q}{2\pi kH}\left[\frac{\langle R^2\rangle}{2R_e^2} + \frac{2kt}{\phi\mu c R_e^2} - \langle \ln R\rangle + \ln R_e - \frac{3}{4}\right] \\ &= P_i - \frac{\mu Q}{2\pi kH}\left[\frac{1}{4} + \frac{2kt}{\phi\mu c R_e^2} - \ln R_e + \frac{1}{2} + \ln R_e - \frac{3}{4}\right]\end{aligned}$$

$$= P_i - \frac{\mu Q}{2\pi kH}\left[\frac{2kt}{\phi\mu c R_e^2}\right]$$

$$= P_i - \frac{Qt}{\phi c \pi R_e^2 H}. \tag{A.76}$$

The derivative of the average reservoir pressure with respect to t is

$$\frac{d\langle P(R,t)\rangle}{dt} = \frac{-Q}{\phi c(\pi R_e^2 H)} = \frac{-Q}{\phi c V}, \tag{A.77}$$

which agrees with Eq. (6.4.16).

If result (A.76) is used to write the pressure distribution (A.69) in terms of the mean reservoir pressure, which is usually denoted by $\bar{P}$ instead of $\langle P\rangle$, we obtain

$$P(R,t) = \bar{P} - \frac{\mu Q}{2\pi kH}\left[\frac{R^2}{2R_e^2} - \ln\left(\frac{R}{R_e}\right) - \frac{3}{4}\right]. \tag{A.78}$$

Evaluating this expression at the wellbore, and again making use of the fact that $R_w \ll R_e$, yields the following relation between the wellbore pressure and the mean reservoir pressure:

$$P_w = \bar{P} - \frac{\mu Q}{2\pi kH}\left[\ln\left(\frac{R_e}{R_w}\right) - \frac{3}{4}\right]. \tag{A.79}$$

This equation leads directly to a commonly-used expression for the *well productivity*, which relates the flow rate to the difference between the mean reservoir pressure and the wellbore pressure (Dake, 1978, p. 145; Matthews and Russell, 1967, p. 13):

$$Q = \frac{2\pi kH(\bar{P} - P_w)}{\mu\left[\ln\left(\frac{R_e}{R_w}\right) - \frac{3}{4}\right]}. \tag{A.80}$$

Problem 6.3. Starting with Eq. (6.3.2), calculate the average pressure in a circular reservoir with a constant pressure outer boundary, during the late-time regime in which all of the exponential terms have died out. Use this result to find an equation for the well productivity, which relates the production rate to the difference between the average reservoir pressure and the pressure at the well.

Solution: After the exponential terms have died out, Eq. (6.3.2) reduces to

$$\Delta P_D(R_D, t_D) = -\ln\left(\frac{R_D}{R_{De}}\right). \tag{A.81}$$

Using Eqs. (6.3.6) and (6.3.8) to revert to dimensional variables, the pressure profile is given by

$$P(R) = P_i + \frac{\mu Q}{2\pi k H}\ln\left(\frac{R}{R_e}\right). \tag{A.82}$$

Following the same procedure as in Problem 6.2, the average reservoir pressure is found to be

$$\begin{aligned}\bar{P} \equiv \langle P(R,t)\rangle &= P_i + \frac{\mu Q}{2\pi k H}\left[\langle \ln R\rangle - \ln R_e\right] \\ &= P_i + \frac{\mu Q}{2\pi k H}\left[\ln R_e - \frac{1}{2} - \ln R_e\right] \\ &= P_i - \left(\frac{1}{2}\right)\frac{\mu Q}{2\pi k H},\end{aligned} \tag{A.83}$$

in which we have used the result $\langle \ln R\rangle = \ln R_e - 0.5$ that was found in the solution to Problem 6.2.

The pressure at the well is found from Eq. (A.82) to be given by

$$P_w = P_i + \frac{\mu Q}{2\pi k H}\ln\left(\frac{R_w}{R_e}\right). \tag{A.84}$$

Using Eqs. (A.83) and (A.84), the difference between the mean reservoir pressure and the well pressure is found to be

$$\begin{aligned}\bar{P} - P_w &= P_i - \frac{\mu Q}{2\pi k H}\left(\frac{1}{2}\right) - P_i - \frac{\mu Q}{2\pi k H}\ln\left(\frac{R_w}{R_e}\right) \\ &= -\frac{\mu Q}{2\pi k H}\left(\frac{1}{2}\right) - \frac{\mu Q}{2\pi k H}\ln\left(\frac{R_w}{R_e}\right) \\ &= \frac{\mu Q}{2\pi k H}\left[\ln\left(\frac{R_e}{R_w}\right) - \frac{1}{2}\right],\end{aligned} \tag{A.85}$$

which can be rearranged to give the following expression for the well productivity:

$$Q = \frac{2\pi kH(\bar{P} - P_w)}{\mu \left[\ln\left(\frac{R_e}{R_w} \right) - \frac{1}{2} \right]}. \tag{A.86}$$

Comparison of this result with the result found in Problem 6.2 shows that the productivity of a well in a circular reservoir with a constant-pressure outer boundary, when defined using the difference between the mean reservoir pressure and the well pressure as the "driving force", differs only slightly from that of a well in a circular reservoir with a closed outer boundary.

Problem 7.1. What is the Laplace transform of the function $f(t) = e^{-\text{at}}$?

Solution: Recall the basic definition given in Eq. (7.1.1)

$$L\{f(t)\} \equiv \widehat{f}(s) \equiv \int_0^\infty f(t)e^{-\text{st}}dt. \tag{A.87}$$

If we insert $f(t) = e^{-\text{at}}$ into Eq. (A.87), we find

$$\begin{aligned} L\{e^{-at}\} &= \int_0^\infty e^{-at}e^{-st}dt = \int_0^\infty e^{-(s+a)t}dt, \\ &= \left. \frac{-e^{-(s+a)t}}{s+a} \right]_0^\infty = \frac{-e^{-\infty} + e^{-0}}{s+a} = \frac{1}{s+a}. \end{aligned} \tag{A.88}$$

Therefore, $L\{e^{-\text{at}}\} = 1/(s+a)$.

We can use this result to find the Laplace transforms of functions that are based on the exponential function. For example,

$$\cosh(\text{at}) = \frac{e^{\text{at}} + e^{-\text{at}}}{2},$$

so:

$$L\{\cosh(\mathrm{at})\} = \tfrac{1}{2}(L\{e^{\mathrm{at}}\} + L\{e^{-\mathrm{at}}\})$$
$$= \frac{1}{2}\left[\frac{1}{s-a} + \frac{1}{s+a}\right] = \frac{1}{2}\left[\frac{s+a}{(s-a)(s+a)} + \frac{s-a}{(s-a)(s+a)}\right]$$
$$= \frac{s}{(s-a)(s+a)} = \frac{s}{s^2-a^2}.$$

Similarly, we can use the result $L\{e^{-\mathrm{at}}\} = 1/(s+a)$ to find the Laplace transform of functions such as sinh(at), $\sin(\omega t)$, and $\cos(\omega t)$.

Problem 7.2. Starting with the basic definition of the Laplace transform, Eq. (7.1.1), verify Eq. (7.1.16).

Solution: Again, we start with the basic definition, Eq. (7.1.1):

$$L\{f(t)\} \equiv \widehat{f}(s) \equiv \int_0^{\infty} f(t)e^{-\mathrm{st}}dt. \tag{A.89}$$

According to this definition, the Laplace transform of the function $f(at)$ is

$$L\{f(at)\} = \int_0^{\infty} f(at)e^{-\mathrm{st}}dt. \tag{A.90}$$

Make a change of variables by defining $x = at$, in which case $t \to x/a$ and $dt \to dx/a$, and the limits of integration remain as 0 and ∞. This yields

$$L\{f(at)\} = \int_0^{\infty} f(x)e^{-s(x/a)}\frac{dx}{a} = \frac{1}{a}\int_0^{\infty} f(x)e^{-(s/a)x}dx. \tag{A.91}$$

But in the last integral in Eq. (A.91), x is just a dummy variable, so we can replace it with t to get

$$L\{f(at)\} = \frac{1}{a}\int_0^{\infty} f(t)e^{-(s/a)t}dt. \tag{A.92}$$

The integral in Eq. (A.92) is just the Laplace transform of $f(t)$, with the Laplace variable s replaced by s/a. Therefore,

$$L\{f(at)\} = \frac{1}{a}\widehat{f}(s)]_{s\to s/a} = \frac{1}{a}\widehat{f}(s/a). \tag{A.93}$$

Problem 7.3. Using the various general properties of Laplace transforms, derive Eq. (7.1.19), $L\{t^n\} = n!/s^{n+1}$, where n is any non-negative integer.

Solution: Equations (7.1.17) and (7.1.18) already show that this expression is true for $n = 0$ and $n = 1$. We can prove the general formula, if we can show that "if this formula is true for an arbitrary value n, then it is necessarily true for $n + 1$"; this type of reasoning is called "mathematical induction".

Start with the fact that

$$\int_0^t \tau^n d\tau = \frac{t^{n+1}}{n+1}. \tag{A.94}$$

Now recall Eq. (7.1.10):

$$L\left\{\int_0^t f(\tau)d\tau\right\} = \frac{1}{s}L\{f(t)\}. \tag{A.95}$$

Apply this rule, with $f(t) = t^n$:

$$L\left\{\int_0^t \tau^n d\tau\right\} = L\left\{\frac{t^{n+1}}{n+1}\right\} = \frac{1}{s}L\{t^n\},$$

$$\text{i.e.} \quad L\{t^{n+1}\} = \frac{n+1}{s}L\{t^n\}. \tag{A.96}$$

Now assume that $L\{t^n\} = n!/s^{n+1}$ is true for "some" value of n. Inserting this into the RHS of Eq. (A.96) yields

$$L\{t^{n+1}\} = \frac{n+1}{s}\frac{n!}{s^{n+1}} = \frac{(n+1)!}{s^{n+2}}. \tag{A.97}$$

If $L\{t^n\} = n!/s^{n+1}$ is true for n, then Eq. (A.97) show that it is also true for $n+1$. But we know that it is true for $n = 0$ because we know that $L\{t^0\} = L\{1\} = 1/s$. Hence, it is true for $n = 1$, and similarly then for $n = 2$, etc. This completes the proof by induction, showing that $L\{t^n\} = n!/s^{n+1}$ for all non-negative integers n.

Problem 7.4. Following the steps that were taken in Section 7.2, use Laplace transforms to solve the problem of linear flow into a hydraulic fracture with *constant pressure* in the fracture:

$$\text{PDE: } \frac{1}{D}\frac{dP}{dt} = \frac{d^2P}{dz^2}, \tag{i}$$

$$\text{IC: } P(z, t = 0) = P_i, \tag{ii}$$

$$\text{far-field BC: } P(z \to \infty, t) = P_i, \tag{iii}$$

$$\text{fracture BC: } P(z = 0, t) = P_f. \tag{iv}$$

First, find the pressure function in the Laplace domain, $\widehat{P}(z, s)$. Next, find an expression for the flow rate into the fracture, in the Laplace domain; call it $\widehat{Q}_f(s)$. Lastly, invert $\widehat{Q}_f(s)$ to find the flow rate into the fracture as a function of time, $Q_f(t)$.

Solution: First, define the Laplace transform of $P(z, t)$:

$$\widehat{P}(z, s) \equiv \int_0^\infty P(z, t)e^{-st}dt. \tag{A.98}$$

Now, take the Laplace transform of *both sides* of the governing PDE, Eq. (i). Using Eq. (7.1.7), and initial condition (ii), the left side of Eq. (i) is transformed as follows:

$$L\left\{\frac{dP}{dt}\right\} = sL\{P(z, t)\} - P(z, t = 0) = s\,\widehat{P}(z, s) - P_i. \tag{A.99}$$

Applying rule (7.1.27) twice, the Laplace transform of the right side of Eq. (i) is

$$L\left\{D\frac{d^2P}{dz^2}\right\} = D\frac{d^2}{dz^2}[L\{P(z, t)\}] = D\frac{d^2\,\widehat{P}(z, s)}{dz^2}. \tag{A.100}$$

So, the transformed representation of Eq. (i) is the following ODE:

$$D\frac{d^2\,\widehat{P}(z, s)}{dz^2} - s\,\widehat{P}(z, s) = -P_i. \tag{A.101}$$

The general solution to Eq. (A.101) is, as in Section 7.2,

$$\widehat{P}(z, s) = Ae^{z\sqrt{s/D}} + Be^{-z\sqrt{s/D}} + \frac{P_i}{s}, \tag{A.102}$$

where A and B are arbitrary constants. The values of A and B are found by satisfying the boundary conditions. Let us start with the Laplace transform of the far-field BC, Eq. (iii). First, for the left-hand side,

$$\begin{aligned} L\{P(z=\infty,t)\} &= L\{\lim_{z\to\infty} P(z,t)\} = \int_0^\infty \lim_{z\to\infty} P(z,t)e^{-\mathrm{st}}dt \\ &= \lim_{z\to\infty} \int_0^\infty P(z,t)e^{-\mathrm{st}}dt \\ &= \lim_{z\to\infty} \widehat{P}(z,s) = \widehat{P}(z\to\infty,s). \end{aligned} \tag{A.103}$$

Now, take the Laplace transform of the right side of Eq. (iii):

$$L\{P_i\} = L\{P_i \cdot 1\} = P_i L\{1\} = \frac{P_i}{s}. \tag{A.104}$$

So, the far-field BC for the transformed pressure function $\widehat{P}(z,s)$ is

$$\widehat{P}(z\to\infty,s) = \frac{P_i}{s}. \tag{A.105}$$

By the same method, the transformed version of Eq. (iv) is

$$\widehat{P}(z=0,s) = \frac{P_f}{s}. \tag{A.106}$$

We now use the boundary conditions (A.105–A.106) to find the constants A and B in the solution (A.102). The far-field boundary condition, Eq. (A.105), shows that A must be zero. The fracture BC, Eq. (A.106), implies that

$$\begin{aligned} \widehat{P}(z=0,s) &= B + \frac{P_i}{s} = \frac{P_f}{s}, \\ \to \quad B &= \frac{P_f - P_i}{s}. \end{aligned} \tag{A.107}$$

The solution to this problem in the Laplace domain is therefore

$$\widehat{P}(z,s) = \frac{(P_f - P_i)}{s} e^{-z\sqrt{s/D}} + \frac{P_i}{s}. \tag{A.108}$$

Next, to find an expression for the flow rate into the fracture in the Laplace domain, we start by applying Darcy's law at the fracture in the *time* domain

$$Q_f(t) = \frac{kA}{\mu}\frac{dP(z=0,t)}{dz}. \tag{A.109}$$

We now take the Laplace transform of both sides of Eq. (A.109). First, on the left side, by definition

$$L\{Q_f(t)\} = \widehat{Q}_f(s). \tag{A.110}$$

Now transform the right side of Eq. (A.109), using rule (7.1.27):

$$L\left\{\frac{kA}{\mu}\frac{dP(z=0,t)}{dz}\right\} = \frac{kA}{\mu}L\left\{\frac{dP(z=0,t)}{dz}\right\} = \frac{kA}{\mu}\frac{d\widehat{P}(z=0,s)}{dz}. \tag{A.111}$$

So, the transformed version of Eq. (A.109) is

$$\widehat{Q}_f(s) = \frac{kA}{\mu}\frac{d\widehat{P}(z=0,s)}{dz}. \tag{A.112}$$

From Eq. (A.108), the derivative that appears in Eq. (A.112) is

$$\frac{d\widehat{P}(z,s)}{dz} = -\sqrt{s/D}\frac{(P_f - P_i)}{s}e^{-z\sqrt{s/D}} = \frac{(P_i - P_f)}{\sqrt{sD}}e^{-z\sqrt{s/D}},$$

$$\rightarrow \quad \frac{d\widehat{P}(z=0,s)}{dz} = \frac{(P_i - P_f)}{\sqrt{D}\sqrt{s}}. \tag{A.113}$$

Combining Eqs. (A.112) and (A.113) gives

$$\widehat{Q}_f(s) = \frac{kA(P_i - P_f)}{\mu\sqrt{D}\sqrt{s}}. \tag{A.114}$$

Finally, we must invert $\widehat{Q}_f(s)$ to find the flow rate into the fracture as a function of time, $Q_f(t)$. Using Eq. (7.1.23) for the inverse

Laplace transform of $s^{-1/2}$, we find:

$$Q_f(t) = L^{-1}\{\widehat{Q}_f(s)\} = L^{-1}\left\{\frac{kA(P_i - P_f)}{\mu\sqrt{D}\sqrt{s}}\right\}$$

$$= \frac{kA(P_i - P_f)}{\mu\sqrt{D}} L^{-1}\{s^{-1/2}\} = \frac{kA(P_i - P_f)}{\mu\sqrt{D\pi t}}. \qquad \text{(A.115)}$$

Now recall that $D = k/\phi\mu c$ to re-write Eq. (A.115) in the form

$$Q_f(t) = \frac{kA(P_i - P_f)}{\mu\sqrt{k\pi t/\phi\mu c}} = A(P_i - P_f)\sqrt{\frac{\phi ck}{\pi\mu t}}. \qquad \text{(A.116)}$$

So, if the pressure in the fracture is constant, the flow rate will drop off as $t^{-1/2}$.

The *cumulative* flow from time 0 until time t is found by integrating the instantaneous flow rate:

$$\int_0^t Q_f(\tau)d\tau = \int_0^t A(P_i - P_f)\sqrt{\frac{\phi ck}{\pi\mu\tau}}d\tau = 2A(P_i - P_f)\sqrt{\frac{\phi ckt}{\pi\mu}}. \qquad \text{(A.117)}$$

The cumulative flow rate therefore increases as $t^{1/2}$.

Problem 8.1. Without looking at the paper by Warren and Root, describe and sketch the way that the drawdown curve in Figure 8.1 would change if (a) the storativity ratio ω increased (or decreased) by a factor of 10, or (b) the transmissivity ratio λ increased (or decreased) by a factor of 10.

Solution: According to Eq. (8.3.4), the early-time straight line is given by

$$\Delta P_{\text{Dw}} = \frac{1}{2}(\ln t_D - \ln\omega + 0.8091). \qquad \text{(A.118)}$$

Therefore, an increase in ω by a factor of 10 would cause this line to move down by an amount $(1/2)\ln\omega = (1/2)\ln(10) = 1.151$ on the semi-log plot. Conversely, decreasing ω by a factor of 10 would cause this line to move upwards by 1.151. The late-time semi-log straight line is given by Eq. (8.3.6):

$$\Delta P_{\text{Dw}} = \frac{1}{2}(t_D + 0.8091). \tag{A.119}$$

This equation does not contain ω, so the late-time straight line will be unaffected by a change in ω. According to Eq. (8.3.12), the intersection time t_{D2} does not depend on ω. So, if ω changes by a factor of 10, and λ does not change, the drawdown curves will look as Figure A.11.

The transmissivity ratio λ does not appear in either (8.3.4) or (8.3.6), so the two semi-log straight lines will not move if λ changes. To isolate the effect of λ, we need to find a property of the drawdown curve that depends on λ but not on ω. Equation (8.3.12) shows that the time at which the horizontal asymptote intersects the late-time straight line is given by

$$t_{D2} = \frac{1}{\gamma\lambda}, \quad \text{or} \quad \ln t_{D2} = -\ln\gamma - \ln\lambda. \tag{A.120}$$

Hence, an increase in λ by a factor of 10 would cause $\ln t_{D2}$ to move to the *left* by an amount $\ln 10 = 2.303$. Conversely, decreasing λ by a factor of 10 would cause $\ln t_{D2}$ to move to the *right* by an amount $\ln 10 = 2.303$.

The drawdowns for the pair of values $\{\omega, \lambda\}$, and also for the values $\{\omega, 10\lambda\}$ and $\{\omega, 0.1\lambda\}$ are shown in Figure A.12. Increased values of λ correspond to larger values of the matrix permeability.

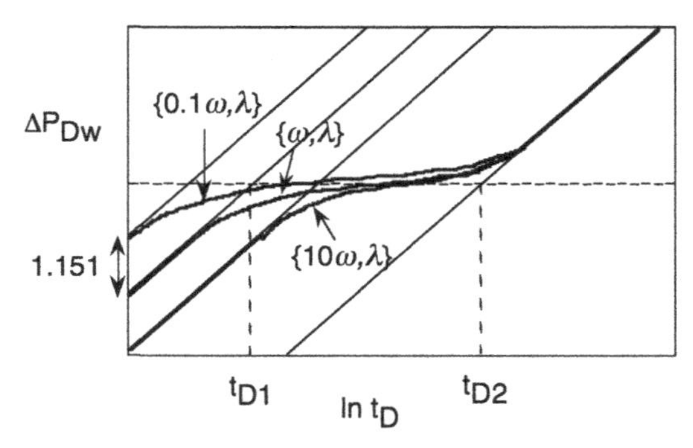

Figure A.11. Schematic graph of the drawdown curves for a well in a dual-porosity reservoir, for different values of the storativity ratio, ω.

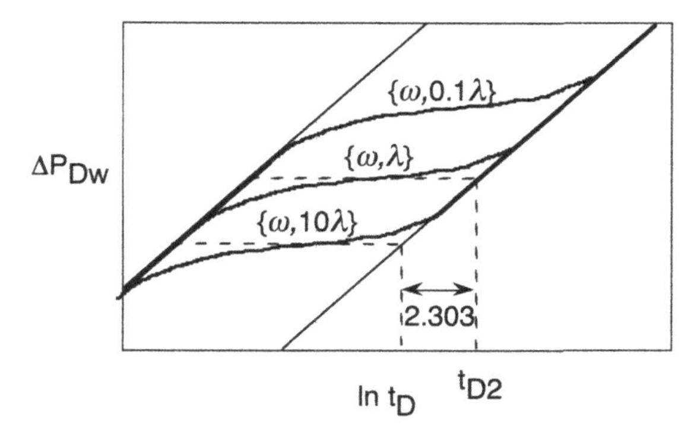

Figure A.12. Schematic graph of the drawdown curves for a well in a dual-porosity reservoir, for different values of the transmissivity ratio, λ.

Hence, it makes sense that if λ increases, the drawdown curve reaches the quasi-steady state (second) straight line in a shorter period of time.

Problem 8.2. By examining Eq. (8.3.2), and making use of either Eq. (2.1.22) or Table 2.1, derive an expression for the time that must elapse in order for the approximation (8.3.6) to be accurate to within about 1%. Your answer should be expressed in terms of the parameter λ.

Solution: For this approximation to hold, the two Ei terms in Eq. (8.3.2) must be negligible compared to the two terms remaining in Eq. (8.3.6). These equations are repeated below:

$$\Delta P_{\mathrm{Dw}} = \frac{1}{2}\left\{\ln t_D + 0.8091 + Ei\left[\frac{-\lambda t_D}{\omega(1-\omega)}\right] - Ei\left[\frac{-\lambda t_D}{(1-\omega)}\right]\right\}, \tag{A.121}$$

$$\Delta P_{\mathrm{Dw}} = \frac{1}{2}(\ln t_D + 0.80901). \tag{A.122}$$

Since most fractured reservoirs will have $\omega \ll 1$, the "x" value in the first Ei function will be much greater than "x" in the second Ei function. But the magnitude of $Ei(-x)$ decreases rapidly as x increases, so we see that we only need to consider the second Ei term.

Therefore, the condition for the Ei terms to be negligible is

$$\left|Ei\left[\frac{-\lambda t_D}{(1-\omega)}\right]\right| < 0.01\left[\frac{1}{2}(\ln t_D + 0.8091)\right]. \tag{A.123}$$

If $\omega \ll 1$, this is equivalent to

$$|Ei(-\lambda t_D)| < 0.01\left[\frac{1}{2}(\ln t_D + 0.8091)\right]. \tag{A.124}$$

Now note that, since we are considering "large" values of time, the term $\ln t_D$ will be at least as large as the term 0.8091. So the bracketed term on the right will be at least as large as about 1. Hence, a sufficient condition for inequality (A.124) to be true is

$$|Ei(-\lambda t_D)| < 0.01. \tag{A.125}$$

If we look at Table 2.1, we see that Eq. (A.125) will hold if the term $\ln t_D$ is greater than about 3. So, the criterion for Eq. (8.3.2) to reduce to Eq. (8.3.6), which is to say, the criterion for the reservoir to behave like a single-porosity reservoir with a storativity equal to the combined storativity of the fractures and matrix blocks, is

$$t_D > 3/\lambda. \tag{A.126}$$

If we look at Figure 5 of Warren and Root (1963), and consider, say, the drawdown curves for $\lambda = 0.005$, we see that they do indeed reach the asymptotic curve (8.3.6) when t_D is about 600, which is exactly consistent with Eq. (A.126). Note also that, according to (8.3.12), $t_{D2} = 1/\gamma\lambda = 0.56/\lambda$. So, result (A.126) is also consistent with Figure 8.3.1, which shows that the drawdown curves reach the second straight-line asymptote somewhat after $t_D = t_{D2}$.

Note 1: The above manipulations were very "approximate", but we somehow arrived at the correct answer. This is due to the fact that the Ei function goes to zero *very* rapidly as x increases. Specifically, for large x, it can be shown from Eq. (2.1.22) that $-Ei(-x) \approx e^{-x}/x$. So, any error in the assumed critical value of $Ei(-x)$ will lead to a very small error in the critical value of x.

Note 2: Mathematically, we always work with the natural log function, $\ln t_D$. In the schematic plots shown above, we used $\ln t_D$, to keep the discussion simple. But more commonly, as in Figure 5 of Warren and Root, "log base 10 of $\ln t_D$", i.e. "$\log t_D$", is used.

Problem 9.1. This problem involves a simplified model for production of gas from a hydraulically fractured shale gas reservoir.

Combining Eqs. (8.1.2) and (8.1.7) leads to the following ODE the governs the mean pressure in a matrix block in a fractured reservoir:

$$\frac{d\bar{P}_m}{dt} = \frac{-\alpha k_m}{\phi_m \mu c_m}(\bar{P}_m - P_f). \tag{A.127}$$

As mentioned in Section 9.2, in a gas reservoir the fluid compressibility term will usually greatly exceed the formation compressibility term, and so the total compressibility of the gas-filled matrix block can be approximated by $c_m = c_{\text{gas}}$. Furthermore, if we approximate the behaviour of the gas as an ideal gas, then according to Eq. (9.2.3), we can say that $c_m = c_{\text{gas}} = 1/P_m$.

(a) Imagine that a matrix block in a gas reservoir is initially at some pressure P_i, after which the pressure in the surrounding fractures is lowered to some value $P_f < P_i$, and then held at that pressure. Approximate the term $c_m = 1/P_m$ by its *initial* value $1/P_i$, and then integrate Eq. (A.127) to find the mean pressure in the matrix block, as a function of time. Then use Eq. (8.1.2) to find an expression for the rate of gas production from the matrix block as a function of time.

(b) Noting that e^{-x} will be < 0.01 when $x > 5$ (roughly), derive an expression for the time at which the production rate of gas from the matrix block has diminished to 1% of its initial value.

(c) When a shale gas reservoir is hydraulically fractured, the interaction between the hydraulically induced fractures and the pre-existing microfractures and bedding planes will create a network of fractures around the wellbore. Imagine that this fracture network breaks up the region around the wellbore into a collection of blocks that can be approximated by cubes of size L. Assuming plausible values such as $L = 1\,\text{m}$, $\phi_m = 0.1$, $\mu = 1 \times 10^{-5}\,\text{Pa s}$, $k_m = 10^{-21}\,\text{m}^2$ and $P_i = 20\,\text{MPa}$, how long will it take for the production rate to drop to 1% of its initial value?

Solution: With the fluid pressure in the fracture, P_f, held constant, the general solution to Eq. (A.127) is readily found to be

$$\bar{P}_m = P_f + A \exp\left(\frac{-\alpha k_m t}{\phi_m \mu c_m}\right), \tag{A.128}$$

where A is an arbitrary constant. Imposing the initial condition that $\bar{P}_m = P_i$ when $t = 0$ shows that $A = P_i - P_f$. Hence, the mean pressure in the matrix block is given by

$$\bar{P}_m = P_f + (P_i - P_f) \exp\left(\frac{-\alpha k_m t}{\phi_m \mu c_m}\right). \tag{A.129}$$

Combining this result with Eq. (8.1.2) shows that the flow rate out of the matrix block is given by

$$q_{\text{mf}} = \frac{\alpha k_m (P_i - P_f)}{\mu} \exp\left(\frac{-\alpha k_m t}{\phi_m \mu c_m}\right). \tag{A.130}$$

This flow rate will have decreased to 1% of its initial value when

$$\frac{\alpha k_m t}{\phi_m \mu c_m} = 5, \quad \text{or} \quad t = \frac{5\phi_m \mu c_m}{\alpha k_m}. \tag{A.131}$$

If the matrix blocks are roughly cubical, of size L, then Eq. (8.1.3) shows that $\alpha = 3\pi^2/L^2$, and so the "depletion time scale" will be given by

$$t_{\text{depleted}} = \frac{5\phi_m \mu c_m L^2}{3\pi^2 k_m} \approx \frac{0.17\phi_m \mu c_m L^2}{k_m}. \tag{A.132}$$

Using the values $L = 1\,\text{m}$, $\phi_m = 0.1$, $\mu = 1 \times 10^{-5}\,\text{Pa s}$, $k_m = 10^{-21}\,\text{m}^2$ and $c_m = 1/P_i = 5 \times 10^{-8}\,/\text{Pa}$, Eq. (A.132) predicts that production will essentially "die out" after about 10^7 sec, or about 100 days. This very rough model provides an explanation of why the production rate of a shale gas well often drops off dramatically after a few months.

Nomenclature

A = cross-sectional area normal to flow (m^2)
A = Drainage area of reservoir (m^2); Chapter 6 only
C_A = Dietz shape factor, Eq. (6.5.2) (-); Chapter 6 only
C_s = wellbore storage coefficient, $= V_w c_f$ (m^3/Pa); Chapter 5 only
C_D = dimensionless wellbore storage coefficient, Eq. (5.3.7) (-); Chapter 5 only
c = compressibility (1/Pa)
c_f = fluid compressibility (1/Pa)
c_t = total compressibility, $= c_f + c_\phi$ (1/Pa)
c_ϕ = formation (pore) compressibility (1/Pa)
D = non-Darcy skin coefficient, Eq. (9.4.11) (s/m^3); Chapter 9 only
D_H = hydraulic diffusivity, $= k/\phi\mu c_t$ (m^2/s)
d = pore diameter (m)
d = distance from well to fault (m); Chapter 4 only
Ei = Exponential integral function, Eq. (2.1.21) (-)
g = gravitational acceleration (m/s^2)
H = reservoir thickness (m)
J_0 = Bessel function of first kind, order zero, Eq. (6.2.37) (-)
J_1 = Bessel function of first kind, order one, Eq. (6.3.5) (-)
k = permeability (m^2)
k_B = Boltzmann constant (Pa m^3/°K); Chapter 9 only
k_f = permeability of fracture network (m^2); Chapter 8 only

k_m = permeability of matrix blocks (m^2); Chapter 8 only
k_R = permeability of undamaged reservoir (m^2); Chapter 5 only
k_{ro} = relative permeability to oil (-); Chapter 1 only
k_{rw} = relative permeability to water (-); Chapter 1 only
k_s = permeability of skin region around well (m^2); Chapter 5 only
L = length of sample (m); Chapter 1 only
L = Laplace transform operator (s); Chapter 7 only
L = length of hydraulic fracture (m); Chapter 7 only
M = generic symbol for a mathematical operator (-); Chapter 3 only
m = mass of fluid inside a region of porous rock (kg); Chapter 1 only
m = magnitude of slope of drawdown *vs.* $\ln t$, Eq. (2.6.3) (Pa); Chapter 2 only
m = real gas pseudopressure, Eq. (9.3.4) (Pa m^2/s); Chapter 9 only
P = pressure (Pa)
P_c = corrected pressure, Eq. (1.2.3) (Pa); Chapter 1 only
P_c = capillary pressure, Eq. (1.8.4) (Pa); Chapter 1 only
P_D = dimensionless pressure = $(P_i - P)/(P_i - P_w)$ (-); Chapter 6 only
P_{Df} = dimensionless pressure in fractures, Eq. (8.2.2) (-); Chapter 8 only
P_{Dm} = dimensionless pressure in matrix blocks, Eq. (8.2.3) (-); Chapter 8 only
P_D^s = steady-state component of dimensionless pressure (-); Chapter 6 only
P_f = pressure in fractures (Pa); Chapter 8 only
P_i = initial reservoir pressure (Pa)
P_m = mean pressure during well test in a gas reservoir, Eq. (9.2.7) (Pa); Chapter 9 only
$\bar{P}_m$ = average pressure in matrix block (Pa); Chapter 8 only
$\widehat{P}$ = Laplace transform of P, Eq. (7.1.2) (Pa s); Chapter 7 only

P^* = characteristic Klinkenberg pressure, Eqs. (9.5.4, 9.5.5) (Pa); Chapter 9 only
ΔP_D = dimensionless drawdown, $= 2\pi kH(P_i - P)/\mu Q$ (-)
ΔP_{Dw} = dimensionless drawdown at the well, $= 2\pi kH(P_i - P_w)/\mu Q$ (-)
ΔP_Q = drawdown per unit of flow rate (Pa s/m^3)
ΔP_s = excess pressure drop in skin region, $= \mu Qs/2\pi k_R H$ (Pa); Chapter 5 only
p_D = transient component of dimensionless pressure (-); Chapter 6 only
Q = volumetric flow rate (m^3/s)
Q_{sf} = sandface flow rate (from reservoir into wellbore) (m^3/s); Chapter 5 only
Q_{wh} = wellhead flow rate (m^3/s); Chapter 5 only
Q^* = total volume injected into well, $= Q\delta t$ (m^3); Chapter 2 only
Q_* = constant (m^3/s); Problem 3.2 only
q = fluid flux, $= Q/A$ (m/s)
q_{mf} = flux from matrix blocks to fractures (1/s); Chapter 8 only
q_o = flux of oil (m/s); Chapter 1 only
q_w = flux of water (m/s); Chapter 1 only
$\mathbf{R}$ = gas constant, Eq. (9.2.1) (Pa m^3/kg°K); Chapter 9 only
R = radial distance from centre of well (m)
R_D = dimensionless radius, $= R/R_w$ (-)
R_{De} dimensionless size of reservoir, $= R_e/R_w$ (-); Chapter 6 only
R_e = outer radius of circular reservoir (m); Chapter 6 only
R_o = outer radius of circular reservoir (m); Chapter 1 and Chapter 5 only
R_s = radius of skin zone (m); Chapter 5 only
R_w = wellbore radius (m)
Re = Reynolds number, Eq. (9.4.1) (-); Chapter 9 only
s = skin factor, Eq. (5.1.7) (-); Chapter 5 only
s = Laplace transform variable (1/s); Chapter 7 only
T = absolute temperature (°K); Chapter 9 only
t = time (s)

t_D = dimensionless time, $= kt/\phi\mu c R^2$ (-)
t_D = dimensionless time, Eq. (8.2.1) (-); Chapter 8 only
$t_{\rm DA}$ = dimensionless time based on the drainage area, Eq. (6.5.3) (-); Chapter 6 only
$t_{\rm Dw}$ = dimensionless time at the well, $= kt/\phi\mu c R_w^2$ (-)
t_H = Horner time, $= (t + \Delta t)/\Delta t$ (-); Chapter 3 only
t_* = constant (s); Problem 3.2 only
Δt = duration of shut-in time during a pressure buildup test (s); Chapter 3 only
v = fluid particle velocity, Eq. (9.4.5) (m/s)
x = Cartesian coordinate (m)
x = temporary variable, $= \lambda R_D$ (-); Chapter 6 only
Y_0 = Bessel function of second kind, order zero, Eq. (6.2.39) (-)
Y_1 = Bessel function of second kind, order one, Eq. (6.3.5) (-)
y = temporary variable, $= \eta(dP/d\eta)$ (Pa); Chapter 2 only
z = vertical coordinate (measured downwards) (m); Chapter 1 only
z = distance from hydraulic fracture (m); Chapter 7 only
z = gas deviation factor, Eq. (9.3.1) (-); Chapter 9 only
z_o = reference depth for computing the corrected pressure (m); Chapter 1 only
α = matrix block shape factor, Eq. (8.1.2) ($1/\text{m}^2$); Chapter 8 only
β = Forchheimer coefficient, Eq. (9.4.9) (Pa s^2/kg); Chapter 9 only
ϕ = porosity (-)
γ = Euler's number, 1.781 (-) NB: some books/papers define $\gamma = \ln(1.781) = 0.5772$
η = Boltzmann variable, $= \mu c R^2/kt$ (-)
λ = eigenvalue (-); Chapter 6 only
λ = transmissivity ratio, Eq. (8.2.12) (-); Chapter 8 only
λ = mean free path of gas molecule, Eq. (9.5.1) (m); Chapter 9 only
μ = viscosity (Pa s)
μ_w = viscosity of water (Pa s); Chapter 1 only

μ_o = viscosity of oil (Pa s); Chapter 1 only
ρ = density (kg/m^3)
σ = effective molecular diameter (m); Chapter 9 only
τ = dummy time-like variable used in integrals (s)
ω = storativity ratio, Eq. (8.2.11) (-); Chapter 8 only

References

Agarwal, R.G., Al-Hussainy, R. and Ramey, H.J. (1970). An investigation of wellbore storage and skin effect in unsteady liquid flow: I. Analytical treatment, *Society of Petroleum Engineers Journal*, **10**(3), 279–290.

Barenblatt, G.I., Zheltov, Y.P. and Kochina, I.N. (1960). Basic concepts in the theory of seepage of homogeneous liquids in fissured rocks, *Journal of Applied Mathematics and Mechanics*, **24**(5), 1286–1303.

Bear, J. (1972). *Dynamics of Fluids in Porous Media*, American Elsevier, New York.

Blunt, M.J. (2017). *Reservoir Engineering: The Imperial College Lectures in Petroleum Engineering*, Vol. 2, World Scientific, London and Singapore.

Bourdet, D. and Gringarten, A.C. (1980). Determination of fissure volume and block size in fractured reservoirs by type-curve analysis. *SPE Annual Fall Technical Conference and Exhibition*, Dallas, Texas, 21–24 September (SPE-9293).

Brigham, W.E., Peden, J.M., Ng, K.F. and O'Neill, N. (1980). The analysis of spherical flow with wellbore storage, *SPE Annual Fall Technical Conference and Exhibition*, Dallas, Texas, 21–24 September (SPE-9294).

Carslaw, H.S. and Jaeger, J.C. (1949). *Operational Methods in Applied Mathematics*, Oxford University Press, Oxford.

Chen, H.K. and Brigham, W.E. (1978). Pressure buildup for a well with storage and skin in a closed square, *Journal of Petroleum Technology*, **36**(1), 141–146.

Chierici, G.L. (1994). *Principles of Petroleum Reservoir Engineering*, Vol. 1, Springer, New York.

Churchill, R.V. (1958). *Operational Mathematics*, McGraw-Hill, New York.

Dake, L.P. (1978). *Fundamentals of Reservoir Engineering*, Elsevier, Amsterdam.

Daltaban, T.S. and Wall, C.G. (1998). *Fundamental and Applied Pressure Analysis*, Imperial College Press, London.

Darcy, H. (1856). *Les Fontaines Publiques de la Ville de Dijon* (The public fountains of the City of Dijon), Dalmont, Paris.

de Marsily, G. (1986). *Quantitative Hydrogeology*, Academic Press, San Diego.

Dietz, D.N. (1965). Determination of average reservoir pressure from build-up surveys, *Journal of Petroleum Technology*, **23**(8), 955–959.

Duhamel, J.N.C. (1833). Sur la méthode générale relative au mouvement de la chaleur dans les corps solides plongés dans des milieux dont la température varie avec le temps (On the general method regarding the movement of heat in a body immersed in a medium whose temperature varies with time), *Journal de l'École Polytechnique*, **14**(22), 20–77.

Dupuit, J. (1857). Mouvement de l'eau a travers le terrains permeables (Movement of water through permeable formations), *Comptes Rendus de l'Académie des Sciences*, **45**, 92–96.

Earlougher, R.C. (1977). *Advances in Well Test Analysis*, Society of Petroleum Engineers, Dallas.

Gringarten, A.C. (1984). Interpretation of tests in fissured and multilayered reservoirs with double-porosity behavior: Theory and practice, *Journal of Petroleum Technology*, **36**(4), 549–564.

Hirschfelder, J.O., Curtiss, C.F. and Bird, R.B. (1954). *Molecular Theory of Gases and Liquids*, Wiley, New York.

Jaeger, J.C., Cook, N.G.W. and Zimmerman, R.W. (2007). *Fundamentals of Rock Mechanics*, 4th edition, Wiley-Blackwell, Oxford.

Joseph, J.A. and Koederitz, L.F. (1985). Unsteady-state spherical flow with storage and skin, *Society of Petroleum Engineers Journal*, **25**(6), 804–822.

Kazemi, H. (1969). Pressure transient analysis of naturally fractured reservoirs with uniform fracture distribution, *Society of Petroleum Engineers Journal*, **9**(4), 451–462.

Kazemi, H., Merrill, L.S., Porterfield, K.L. and Zeman, P.R. (1976). Numerical simulation of water-oil flow in naturally fractured reservoirs, *Society of Petroleum Engineers Journal*, **16**(6), 317–326.

Klinkenberg, L.J. (1941). The Permeability of Porous Media to Liquids and Gases, in *Drilling and Production Practices*, American Petroleum Institute, Washington, D.C., pp. 200–213.

Matthews, C.S. and Russell, D.G. (1967). *Pressure Buildup and Flow Tests in Wells*, Society of Petroleum Engineers, Dallas.

Muskat, M. (1937). *The Flow of Homogeneous Fluids through Porous Media*, McGraw-Hill, New York.

Quintard, M. and Whitaker, S. (1996). Transport in chemically and mechanically heterogeneous porous media: 2, Comparison with numerical experiments for slightly compressible single-phase flow, *Advances in Water Resources*, **19**(1), 49–60.

Stanislav, J.F. and Kabir, C.S. (1990). *Pressure Transient Analysis*, Prentice-Hall, Englewood Cliffs, New Jersey.

Stehfest, H. (1968). Algorithm 368: Numerical inversions of Laplace transforms, *Communications of the Association for Computing Machinery*, **13**(1), 47–49.

Stewart, G. (2011). *Well Test Design and Analysis*, Pennwell Publishers, Tulsa, Oklahoma.

Streltsova, T.D. (1988). *Well Testing in Heterogeneous Formations*, Wiley, New York.

Theis, C.V. (1935). The relation between the lowering of the piezometric surface and the rate and duration of discharge of a well using groundwater storage, *Transactions American Geophysical Union*, **16**, 519–524.

Thiem, A. (1887). Verfahress für Natürlicher Grundwassergeschwindegkiten (Movement of natural groundwater flow). *Polytechnisches Notizblatt*, **42**, 229.

Tranter, C.J. (1971). *Integral Transforms in Mathematical Physics*, Chapman and Hall, London.

van Everdingen, A.F. and Hurst, W. (1949). The application of the Laplace transformation to flow problems in reservoirs, *Petroleum Transactions, AIME*, **186**, 305–324.

Warren, J.E. and Root, P.J. (1963). The behavior of naturally fractured reservoirs, *Society of Petroleum Engineers Journal*, **3**(3), 245–255.

Watson, G.N. (1944). *A Treatise on the Theory of Bessel Functions*, 2nd edition, Cambridge University Press, Cambridge.

Wattenbarger, R.A. and Ramey, H.J. (1970). An investigation of wellbore storage and skin effect in unsteady liquid flow: ii. Finite difference treatment, *Society of Petroleum Engineers Journal*, **10**(3), 291–297.

Zimmerman, R.W. (1991). *Compressibility of Sandstones*, Elsevier, Amsterdam.

Zimmerman, R.W. (2017a), Introduction to rock properties, in *Topics in Reservoir Management: The Imperial College Lectures in Petroleum Engineering*, Vol. 3, World Scientific, London and Singapore, pp. 1–46.

Zimmerman, R.W. (2017b). Pore volume and porosity changes under uniaxial strain conditions, *Transport in Porous Media*, **119**(2), 481–498.

Zimmerman, R.W., Chen, G., Hadgu, T. and Bodvarsson, G.S. (1993). A numerical dual-porosity model with semianalytical treatment of fracture/matrix flow, *Water Resources Research*, **29**(7), 2127–2137.

Index

D

E

F

G

Printed in the United States
By Bookmasters